This is a volume in
PURE AND APPLIED MATHEMATICS

A Series of Monographs and Textbooks

Editors: SAMUEL EILENBERG AND HYMAN BASS

A list of recent titles in this series appears at the end of this volume.

HOMOLOGY THEORY

An Introduction to Algebraic Topology

homology theory

AN INTRODUCTION
TO ALGEBRAIC TOPOLOGY

JAMES W. VICK

University of Texas

ACADEMIC PRESS   New York   San Francisco   London

A Subsidiary of Harcourt Brace Jovanovich, Publishers

ACADEMIC PRESS, INC.
111 Fifth Avenue, New York, New York 10003

United Kingdom Edition published by
ACADEMIC PRESS, INC. (LONDON) LTD.
24/28 Oval Road, London NW1

LIBRARY OF CONGRESS CATALOG CARD NUMBER: 72-84375

AMS (MOS) 1970 Subject Classifications: 55-01, 55B10, 57-10

PRINTED IN THE UNITED STATES OF AMERICA

to Professor Ed Floyd

contents

preface

During the last twenty-five years the field of algebraic topology has experienced a period of phenomenal growth and development. Along with the increasing number of students and researchers in the field and the expanding areas of knowledge have come new applications of the techniques and results of algebraic topology in other branches of mathematics. As a result there has been a growing demand for an introductory course in algebraic topology for students in algebra, geometry, and analysis, as well as for those planning further work in topology.

This book is designed as a text for such a course as well as a source for individual reading and study. Its purpose is to present as clearly and concisely as possible the basic techniques and applications of homology theory. The subject matter includes singular homology theory, attaching spaces and finite CW complexes, cellular homology, the Eilenberg–Steenrod axioms, cohomology, products, and duality and fixed-point theory for topological manifolds. The treatment is highly intuitive with many figures to increase the geometric understanding. Generalities have been avoided whenever it was felt that they might obscure the essential concepts.

Although the prerequisites are limited to basic algebra (abelian groups) and general topology (compact Hausdorff spaces), a number of the classical applications of algebraic topology are given in the first chapter. Rather than devoting an initial chapter to homological algebra, these concepts have been integrated into the text so that the motivation for the constructions is more apparent. Similarly the exercises have been spread throughout in order to exploit techniques or reinforce concepts.

At the close of the book there are three bibliographical lists. The first includes all works referenced in the text. The second is an extensive list of books and notes in algebraic topology and related fields, and the third is a similar list of survey and expository articles. It was felt that these would best serve the student, teacher, and reader in offering accessible sources for further reading and study.

acknowledgments

The original manuscript for this book was a set of lecture notes from Math 401–402 taught at Princeton University in 1969–1970. However, much of the technique and organization of the first four chapters may be traced to courses in algebraic topology taught by Professor E. E. Floyd at the University of Virginia in 1964–1965 and 1966–1967. The author was one of the fortunate students who have been introduced to the subject by such a masterful teacher. Any compliments that this book may merit should justifiably be directed first to Professor Floyd.

The author wishes to express his appreciation to the students and faculty of Princeton University and the University of Texas who have taken an interest in these notes, contributed to their improvement, and encouraged their publication. The typing of the manuscript by the secretarial staff of the Mathematics Department at the University of Texas was excellent, and particular thanks are due to Diane Schade who typed the majority of it. Many helpful improvements and corrections in the original manuscript were suggested by Professor Peter Landweber.

Finally the author expresses a deep sense of gratitude to his wife and family for their boundless patience and understanding over years during which this book has evolved.

HOMOLOGY THEORY

An Introduction to Algebraic Topology

chapter 1

SINGULAR HOMOLOGY THEORY

The purpose of this chapter is to introduce the singular homology theory of an arbitrary topological space. Following the definitions and a proof of homotopy invariance, the essential computational tool (Theorem 1.14) is stated. Its proof is deferred to Appendix I so that the exposition need not be interrupted by its involved constructions. The Mayer–Vietoris sequence is noted as an immediate corollary of this theorem, and then applied to compute the homology groups of spheres. These results are applied to prove a number of classical theorems: the nonretractibility of a disk onto its boundary, the Brouwer fixed-point theorem, the nonexistence of vector fields on even-dimensional spheres, the Jordan–Brouwer separation theorem and the Brouwer theorem on the invariance of domain.

If x and y are points in R^n, define the *segment* from x to y to be $\{(1 - t)x + ty \mid 0 \leq t \leq 1\}$. A subset $C \subseteq R^n$ is *convex* if, given x and y in C, the segment from x to y lies entirely in C. Note that an arbitrary intersection of convex sets is convex. If $A \subseteq R^n$, the *convex hull* of A is the intersection of all convex sets in R^n which contain A.

A p-simplex s in R^n is the convex hull of a collection of $(p + 1)$ points $\{x_0, \ldots, x_p\}$ in R^n in which $x_1 - x_0, \ldots, x_p - x_0$ form a linearly independent set. Note that this is independent of the designation of which point is x_0.

1

1.1   PROPOSITION   Let $\{x_0, \ldots, x_p\} \subseteq R^n$. Then the following are equivalent:

(a) $x_1 - x_0, \ldots, x_p - x_0$ are linearly independent;
(b) if $\sum s_i x_i = \sum t_i x_i$ and $\sum s_i = \sum t_i$, then $s_i = t_i$ for $i = 0, \ldots, p$.

PROOF   (a) $\Rightarrow$ (b): If $\sum s_i x_i = \sum t_i x_i$ and $\sum s_i = \sum t_i$, then

$$0 = \sum_{i=0}^{p} (s_i - t_i) x_i = \sum_{i=0}^{p} (s_i - t_i) x_i - \left[\sum_{i=0}^{p} (s_i - t_i) \right] x_0$$

$$= \sum_{i=1}^{p} (s_i - t_i)(x_i - x_0).$$

By the linear independence of $x_1 - x_0, \ldots, x_p - x_0$ it follows that $s_i = t_i$ for $i = 1, \ldots, p$. Finally, this implies $s_0 = t_0$ since $\sum s_i = \sum t_i$.

(b) $\Rightarrow$ (a): If $\sum_{i=1}^{p} (t_i)(x_i - x_0) = 0$, then $\sum_{i=1}^{p} t_i x_i = (\sum_{i=1}^{p} t_i) x_0$ and by (b) the coefficients $t_1, \ldots, t_n$ must all be zero. This proves linear independence. $\square$

Let s be a p-simplex in R^n and consider the set of all points of the form $t_0 x_0 + t_1 x_1 + \cdots + t_p x_p$, where $\sum t_i = 1$ and $t_i \geq 0$ for each i. Note that this is the convex hull of the set $\{x_0, \ldots, x_p\}$ and hence from Proposition 1.1 we have the following:

1.2   PROPOSITION   If the p-simplex s is the convex hull of $\{x_0, \ldots, x_p\}$, then every point of s has a distinct unique representation in the form $\sum t_i x_i$, where $t_i \geq 0$ for all i and $\sum t_i = 1$. $\square$

The points x_i are the *vertices* of s. This proposition allows us to associate the points of s with $(p + 1)$-tuples $(t_0, t_1, \ldots, t_p)$ with a suitable choice of the coordinates t_i.

Exercise 1   Let y be a point in s. Then y is a vertex of s if and only if y is not an interior point of any segment lying in s.

If the vertices of s have been given a specific order, then s is an *ordered simplex*. So let s be an ordered simplex with vertices $x_0, x_1, \ldots, x_p$. Define σ_p to be the set of all points $(t_0, t_1, \ldots, t_p) \in R^{p+1}$ with $\sum t_i = 1$ and $t_i \geq 0$ for each i. If a function

$$f: \quad \sigma_p \to s$$

is given by $f(t_0, \ldots, t_p) = \sum t_i x_i$, then f is continuous. Moreover, from the uniqueness of representations and the fact that σ_p and s are compact

Hausdorff spaces it follows that f is a homeomorphism. Thus, each ordered p-simplex is a natural homeomorphic image of σ_p. Note that σ_p is a p-simplex with vertices $x_0' = (1, 0, \ldots, 0)$, $x_1' = (0, 1, \ldots, 0)$, $\ldots$, $x_p' = (0, \ldots, 0, 1)$. σ_p is called the *standard p-simplex with natural ordering*.

Let X be a topological space. A *singular p-simplex* in X is a continuous function

$$\phi: \quad \sigma_p \to X.$$

Note that the singular 0-simplices may be identified with the points of X, the singular 1-simplices with the paths in X, and so forth.

If ϕ is a singular p-simplex and i is an integer with $0 \le i \le p$, define $\partial_i(\phi)$, a singular $(p-1)$-simplex in X, by

$$\partial_i \phi(t_0, \ldots, t_{p-1}) = \phi(t_0, t_1, \ldots, t_{i-1}, 0, t_i, \ldots, t_{p-1}).$$

$\partial_i \phi$ is the *i*th *face* of ϕ.

For example, let ϕ be a singular 2-simplex in X (Figure 1.1). Then, $\partial_1 \phi$ is given by the composition shown in Figure 1.2. That is, to compute $\partial_i \phi$ we embed σ_{p-1} into σ_p opposite the *i*th vertex, using the usual ordering of vertices, and then go into X via ϕ.

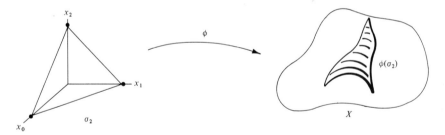

Figure 1.1

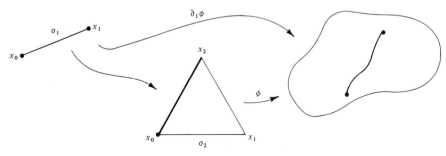

Figure 1.2

If $f: X \to Y$ is a continuous function and ϕ is a singular p-simplex in X, define a singular p-simplex $f_\#(\phi)$ in Y by $f_\#(\phi) = f \circ \phi$. Note that if $g: Y \to W$ is continuous and id: $X \to X$ is the identity map,

$$(g \circ f)_\#(\phi) = g_\#(f_\#(\phi)) \qquad \text{and} \qquad (\text{id})_\#(\phi) = \phi.$$

An abelian group G is *free* if there exists a subset $A \subseteq G$ such that every element g in G has a unique representation

$$g = \sum_{x \in A} n_x \cdot x,$$

where n_x is an integer and equal to zero for all but finitely many x in A. The set A is a *basis* for G.

Given an arbitrary set A we may construct a free abelian group with basis A in the following manner. Let $F(A)$ be the set of all functions f from A into the integers such that $f(x) \neq 0$ for only a finite number of elements of A. Define an operation in $F(A)$ by $(f + g)(x) = f(x) + g(x)$. Then $F(A)$ is an abelian group. For any $a \in A$ define a function f_a in $F(A)$ by

$$f_a(x) = \begin{cases} 1 & \text{if} \quad x = a \\ 0 & \text{otherwise.} \end{cases}$$

Then $\{ f_a \mid a \in A \}$ is a basis for $F(A)$ as a free abelian group. Identifying a with f_a completes the construction.

For example, let $G = \{(n_1, n_2, \ldots) \mid n_i \text{ is an integer, eventually } 0\}$. Then G is an abelian group under coordinatewise addition, and furthermore it is free with basis

$$(1, 0, \ldots), \quad (0, 1, 0, \ldots), \quad (0, 0, 1, 0, \ldots), \quad \ldots.$$

For convenience we say that if $G = 0$, then G is a free abelian group with empty basis.

Note that if G is free abelian with basis A and H is an abelian group, then every function $f: A \to H$ can be uniquely extended to a homomorphism $f: G \to H$.

If X is a topological space define $S_n(X)$ to be the free abelian group whose basis is the set of all singular n-simplices of X. An element of $S_n(X)$ is called a *singular n-chain* of X and has the form

$$\sum_\phi n_\phi \cdot \phi,$$

where n_ϕ is an integer, equal to zero for all but a finite number of ϕ.

Since the ith face operator ∂_i is a function from the set of singular n-simplices to the set of singular $(n-1)$-simplices, there is a unique extension to a homomorphism

$$\partial_i: \quad S_n(X) \to S_{n-1}(X)$$

given by $\partial_i(\sum n_\phi \cdot \phi) = \sum n_\phi \cdot \partial_i\phi$. Define the *boundary operator* to be the homomorphism

$$\partial: \quad S_n(X) \to S_{n-1}(X)$$

given by

$$\partial = \partial_0 - \partial_1 + \partial_2 + \cdots + (-1)^n\partial_n = \sum_{i=0}^{n} (-1)^i\partial_i.$$

1.3 PROPOSITION   The composition $\partial \circ \partial$ in

$$S_n(X) \xrightarrow{\partial} S_{n-1}(X) \xrightarrow{\partial} S_{n-2}(X)$$

is zero.

Exercise 2   Prove Proposition 1.3. □

Geometrically this statement merely says that the boundary of any n-chain is an $(n-1)$-chain having no boundary. It is this basic property which leads to the definition of the homology groups. An element $c \in S_n(X)$ is an n-*cycle* if $\partial(c) = 0$. An element $d \in S_n(X)$ is an n-*boundary* if $d = \partial(e)$ for some $e \in S_{n+1}(X)$. Since ∂ is a homomorphism, its kernel, the set of all n-cycles, is a subgroup of $S_n(X)$ denoted by $Z_n(X)$. Similarly the image of ∂ in $S_n(X)$ is the subgroup $B_n(X)$ of all n-boundaries.

Note that Proposition 1.3 implies that $B_n(X) \subseteq Z_n(X)$ is a subgroup. The quotient group

$$H_n(X) = Z_n(X)/B_n(X)$$

is the nth *singular homology group* of X. The geometric motivation for this algebraic construction is evident; the objects we wish to study are cycles in topological spaces. However, in using singular cycles, the collection of all such is too vast to be effectively studied. The natural approach is then to restrict our attention to equivalence classes of cycles under the relation that two cycles are equivalent if their difference forms a boundary of a chain of one dimension higher.

This algebraic technique is a standard construction in homological algebra. A *graded* (abelian) *group* G is a collection of abelian groups $\{G_i\}$

indexed by the integers with componentwise operation. If G and G' are graded groups, a homomorphism

$$f\colon \ G \to G'$$

is a collection of homomorphisms $\{f_i\}$, where

$$f_i\colon \ G_i \to G'_{i+r}$$

for some fixed integer r. r is then called the *degree* of f. A subgroup $H \subseteq G$ of a graded group is a graded group $\{H_i\}$ where H_i is a subgroup of G_i. The quotient group G/H is the graded group $\{G_i/H_i\}$.

A *chain complex* is a sequence of abelian groups and homomorphisms

$$\cdots \xrightarrow{\partial_{n+1}} C_n \xrightarrow{\partial_n} C_{n-1} \xrightarrow{\partial_{n-1}} \cdots$$

in which the composition $\partial_{n-1} \circ \partial_n = 0$ for each n. Equivalently a chain complex is a graded group $C = \{C_i\}$ together with a homomorphism $\partial\colon C \to C$ of degree -1 such that $\partial \circ \partial = 0$. If C and C' are chain complexes with boundary operators ∂ and ∂', a chain map from C to C' is a homomorphism

$$\Phi\colon \ C \to C'$$

of degree zero such that $\partial' \circ \Phi_n = \Phi_{n-1} \circ \partial$ for each n. (Note that the requirement that Φ have degree zero is unnecessary. It is stated here only as a convenience since all chain maps we will consider have this property.) Denoting by $Z_*(C)$ and $B_*(C)$ the kernel and image of ∂, respectively, the homology of C is the graded group

$$H_*(C) = Z_*(C)/B_*(C).$$

Note that if Φ is a chain map,

$$\Phi(Z_*(C)) \subseteq Z_*(C') \qquad \text{and} \qquad \Phi(B_*(C)) \subseteq B_*(C').$$

Therefore, Φ induces a homomorphism on homology groups

$$\Phi_*\colon \ H_*(C) \to H_*(C').$$

In this sense the graded group $S_*(X) = \{S_i(X)\}$ becomes a chain complex under the boundary operator ∂, so that the homology group of X is the homology of this chain complex. If $f\colon X \to Y$ is a continuous function and ϕ is a singular n-simplex in X, there is the singular n-simplex $f_\#(\phi) = f \circ \phi$

in Y. This extends uniquely to a homorphism

$$f_\#: \quad S_n(X) \to S_n(Y) \qquad \text{for each} \quad n.$$

To show that $f_\#$ is a chain map from $S_*(X)$ to $S_*(Y)$ it must be checked that the following rectangle commutes:

$$
\begin{array}{ccc}
S_n(X) & \xrightarrow{\ f_\#\ } & S_n(Y) \\
\downarrow{\scriptstyle \partial} & & \downarrow{\scriptstyle \partial} \\
S_{n-1}(X) & \xrightarrow{\ f_\#\ } & S_{n-1}(Y)
\end{array}
$$

First note that it is sufficient to check that this is true on singular n-simplices ϕ, and second, observe that it is sufficient to show $\partial_i f_\#(\phi) = f_\# \partial_i(\phi)$. Now

$$f_\# \partial_i(\phi)(t_0, \ldots, t_{n-1}) = f(\phi(t_0, \ldots, t_{i-1}, 0, t_i, \ldots, t_{n-1}))$$

and

$$\partial_i f_\#(\phi)(t_0, \ldots, t_{n-1}) = f_\#(\phi)(t_0, \ldots, t_{i-1}, 0, t_i, \ldots, t_{n-1})$$
$$= f(\phi(t_0, \ldots, t_{i-1}, 0, t_i, \ldots, t_{n-1})).$$

Thus, $f_\#: S_*(X) \to S_*(Y)$ is a chain map and there is induced a homomorphism of degree zero

$$f_*: \quad H_*(X) \to H_*(Y).$$

Note that this is suitably functorial in the sense that for $g: Y \to W$ a continuous function and $\mathrm{id}: X \to X$ the identity, $(g \circ f)_* = g_* \circ f_*$ and id_* is the identity homomorphism.

As a first example take $X = $ point. Then for each $p \geq 0$ there exists a unique singular p-simplex $\phi_p: \sigma_p \to X$. Note further that for $p > 0$, $\partial_i \phi_p = \phi_{p-1}$. So consider the chain complex

$$\cdots \to S_2(\mathrm{pt}) \to S_1(\mathrm{pt}) \to S_0(\mathrm{pt}) \to 0.$$

Each $S_n(\mathrm{pt})$ is an infinite cyclic group generated by ϕ_n. The boundary operator is given by

$$\partial \phi_n = \sum_{i=0}^{n} (-1)^i \partial_i \phi_n = \sum_{i=0}^{n} (-1)^i \phi_{n-1}.$$

Thus, $\partial \phi_{2n-1} = 0$ and $\partial \phi_{2n} = \phi_{2n-1}$ for $n > 0$. Applying this to the chain complex it is evident that

$$Z_n(\text{pt}) = B_n(\text{pt}) \qquad \text{for} \quad n > 0.$$

However, $Z_0(\text{pt}) = S_0(\text{pt})$ is infinite cyclic, whereas $B_0(\text{pt}) = 0$. Therefore, we conclude that the homology groups of a point are given by

$$H_n(\text{pt}) = \begin{cases} Z & \text{if} \quad n = 0 \\ 0 & \text{if} \quad n > 0. \end{cases}$$

A space X is *pathwise connected* if given $x, y \in X$, there is a continuous function

$$\psi: \quad [0, 1] \to X$$

such that $\psi(0) = x$ and $\psi(1) = y$. Note that instead of $[0, 1]$ we could have used σ_1.

Suppose that X is a pathwise connected space, and consider the portion of the singular chain complex of X given by

$$S_1(X) \xrightarrow{\partial} S_0(X) \to 0.$$

Now $S_0(X) = Z_0(X)$, which may be viewed as the free abelian group generated by the points of X. That is $Z_0(X) = F(X)$. Hence, an element y of $Z_0(X)$ has the form

$$y = \sum_{x \in X} n_x \cdot x,$$

where the n_x are integers, all but finitely many equal to zero.

On the other hand, $S_1(X)$ may be viewed as the free abelian group generated by the set of all paths in X. If the vertices of σ_1 are v_0 and v_1 and ϕ is a singular 1-simplex in X, then

$$\partial \phi = \phi(v_1) - \phi(v_0) \in Z_0(X).$$

Define a homomorphism $\alpha: S_0(X) \to Z$ by $\alpha(\sum n_x \cdot x) = \sum n_x$. Note that if X is nonempty, then α is an epimorphism. Since for any singular 1-simplex ϕ in X, $\alpha(\partial \phi) = \alpha(\phi(v_1) - \phi(v_0)) = 0$, it follows that $B_0(X)$ is contained in the kernel of α.

Conversely, suppose that $n_1 x_1 + \cdots + n_k x_k \in Z_0(X)$ with $\sum n_i = 0$. Pick any point $x \in X$ and note that for each i there is a singular 1-simplex ϕ_i: $\sigma_1 \to X$ with $\partial_0(\phi_i) = x_i$ and $\partial_1(\phi_i) = x$. Taking the singular 1-chain $\sum n_i \phi_i$ in $S_1(X)$ we have $\partial(\sum n_i \phi_i) = \sum n_i x_i - (\sum n_i)x = \sum n_i x_i$. Therefore, the kernel of α is contained in $B_0(X)$. This proves that the kernel of α equals $B_0(X)$ and we conclude the following:

1.4 PROPOSITION If X is a nonempty pathwise-connected space, then

$$H_0(X) \approx Z. \quad \square$$

Let A be a set and suppose that for each $\alpha \in A$ there is given an abelian group G_α. Define an abelian group $\sum_{\alpha \in A} G_\alpha$ as follows: the elements are all functions

$$f: \quad A \rightarrow \bigcup_{\alpha \in A} G_\alpha$$

such that $f(\alpha) \in G_\alpha$ for each α, and $f(\alpha) = 0$ for all but finitely many elements $\alpha \in A$; the operation is defined by $(f + g)(\alpha) = f(\alpha) + g(\alpha)$. Setting $g_\alpha = f(\alpha) \in G_\alpha$ we write $f = (g_\alpha : \alpha \in A)$ and call the g_α the components of f. The group $\sum G_\alpha$ is the *weak direct sum* of the G_α's. If the requirement that $f(\alpha) = 0$ for all but finitely many α is omitted, then the resulting group is the *strong direct sum* or *direct product* of the G_α's, denoted $\Pi_{\alpha \in A} G_\alpha$.

Note that if G is an abelian group and $\{G_\alpha\}_{\alpha \in A}$ is a family of subgroups of G such that $g \in G$ has a unique representation

$$g = \sum_{\alpha \in A} g_\alpha \quad \text{with} \quad g_\alpha \in G_\alpha$$

and $g_\alpha = 0$ for all but finitely many α, then G is isomorphic to $\sum_{\alpha \in A} G_\alpha$.

Now for each $\alpha \in A$ suppose we have a chain complex C^α

$$\cdots \xrightarrow{\partial^\alpha} C_p{}^\alpha \xrightarrow{\partial^\alpha} C_{p-1}^\alpha \xrightarrow{\partial^\alpha} \cdots.$$

Define a chain complex $\sum_{\alpha \in A} C^\alpha$ by taking $(\sum C^\alpha)_p = \sum C_p{}^\alpha$ and setting $\partial(c_\alpha : \alpha \in A) = (\partial^\alpha c_\alpha : \alpha \in A)$.

1.5 LEMMA $H_k(\sum C^\alpha) \approx \sum_\alpha H_k(C^\alpha)$.

PROOF Note that by the definition of the chain complex $\sum C^\alpha$ we have

$$Z_k(\sum C^\alpha) = \sum (Z_k(C^\alpha)) \quad \text{and} \quad B_k(\sum (C^\alpha)) = \sum (B_k(C^\alpha)).$$

Therefore

$$
\begin{aligned}
H_k(\sum C^\alpha) &= Z_k(\sum C^\alpha)/B_k(\sum C^\alpha) \\
&= \sum (Z_k(C^\alpha))/\sum (B_k(C^\alpha)) \\
&\approx \sum (Z_k(C^\alpha)/B_k(C^\alpha)) \\
&= \sum H_k(C^\alpha). \quad \square
\end{aligned}
$$

Let X be a topological space and for $x, y \in X$, set $x \sim y$ if there exists a path in X from x to y. It is evident that $\sim$ is an equivalence relation, that is,

(1) $x \sim x$,
(2) $x \sim y$ and $y \sim z$ implies $x \sim z$,
(3) $x \sim y$ implies $y \sim x$,

for all points x, y, and z in X. Such a relation decomposes X into a collection of subsets, the equivalence classes, where x and y are in the same equivalence class if and only if $x \sim y$. For this specific relation on X the equivalence classes are called the *path components* of X. Note that if $x \in X$ the path component of X containing x is the maximal pathwise-connected subset of X containing x.

1.6 PROPOSITION If X is a space and $\{X_\alpha : \alpha \in A\}$ are the path components of X, then

$$H_k(X) \approx \sum_{\alpha \in A} H_k(X_\alpha).$$

PROOF There is a natural homomorphism

$$\Psi : \sum_{\alpha \in A} S_k(X_\alpha) \to S_k(X)$$

given by

$$\Psi\left(\left(\sum_{\phi_\alpha} n_{\phi,\alpha} \cdot \phi_\alpha\right) : \alpha \in A\right) = \sum_{\alpha \in A} \left(\sum_{\phi_\alpha} n_{\phi,\alpha} \cdot \phi_\alpha\right).$$

Since the groups involved are free abelian, Ψ must be a monomorphism. To observe that Ψ is also an epimorphism, note first that if

$$\phi : \sigma_k \to X$$

is a singular k-simplex, then $\phi(\sigma_k)$ is contained in some X_α because σ_k is pathwise connected. Hence, to any such ϕ there is associated a unique $\phi_\alpha \in S_k(X_\alpha)$ with $\Psi(\phi_\alpha) = \phi$. Therefore, Ψ is an isomorphism for each k. Moreover, Ψ is a chain map between chain complexes so that

$$H_k(X) \approx H_k\left(\sum_{\alpha \in A} S_*(X_\alpha)\right).$$

Finally, it follows from Lemma 1.5 that

$$H_k\left(\sum_{\alpha \in A} S_*(X_\alpha)\right) \approx \sum_{\alpha \in A} H_k(X_\alpha),$$

which completes the proof. □

This proposition establishes the intrinsic "additive" property of singular homology theory. Since the homological properties of a space are completely determined by those of its path components, and the homological properties of any path component are independent of the properties of any other path component, we may restrict our attention to the study of pathwise-connected spaces.

Note that it follows from Propositions 1.6 and 1.4 that $H_0(X)$ is a free abelian group whose basis is in a one-to-one correspondence with the path components of X.

1.7   THEOREM   If $f\colon X \to Y$ is a homeomorphism, then

$$f_*\colon\quad H_p(X) \to H_p(Y)$$

is an isomorphism for each p.

Exercise 3   Prove Theorem 1.7.   □

The fact that this theorem, the topological invariance of the singular homology groups, is quite easy to prove is one of the major advantages of using singular homology theory.

1.8   THEOREM   If X is a convex subset of $\mathbb{R}^n$, then

$$H_p(X) = 0 \qquad \text{for } p > 0.$$

PROOF   Assume $X \neq \emptyset$ and let $x \in X$ and $\phi\colon \sigma_p \to X$ be a singular p-simplex, $p \geq 0$. Then define a singular $(p+1)$-simplex $\theta\colon \sigma_{p+1} \to X$ as follows:

$$\theta(t_0, \dots, t_{p+1}) = \begin{cases} (1-t_0) \cdot \left(\phi\!\left(\dfrac{t_1}{1-t_0}, \dots, \dfrac{t_{p+1}}{1-t_0} \right) \right) + t_0 x & \text{for } t_0 < 1 \\ x & \text{for } t_0 = 1. \end{cases}$$

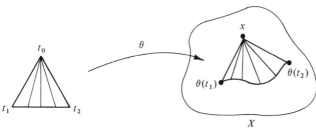

Figure 1.3

That is, we are setting

$$\theta(0, t_1, \ldots, t_{p+1}) = \phi(t_1, \ldots, t_{p+1}) \quad \text{and} \quad \theta(1, 0, \ldots, 0) = x$$

and then taking line segments from t_0 to the face opposite t_0 linearly into the corresponding line segment in X (Figure 1.3). This construction is possible since X is convex.

From its definition θ is continuous except possibly at $(1, 0, \ldots, 0)$. To check continuity there we must show that

$$\lim_{t_0 \to 1} \| \theta(t_0, \ldots, t_{p+1}) - x \| = 0.$$

Now

$$\lim_{t_0 \to 1} \| \theta(t_0, \ldots, t_{p+1}) - x \|$$

$$= \lim_{t_0 \to 1} \left\| (1 - t_0)\left(\phi\left(\frac{t_1}{1 - t_0}, \ldots, \frac{t_{p+1}}{1 - t_0}\right)\right) - (1 - t_0)x \right\|$$

$$\leq \lim_{t_0 \to 1} (1 - t_0)\left(\left\|\phi\left(\frac{t_1}{1 - t_0}, \ldots, \frac{t_{p+1}}{1 - t_0}\right)\right\| + \| x \|\right).$$

Since $\phi(\sigma_p)$ is compact, $(\| \phi(t_1/(1 - t_0), \ldots, t_{p+1}/(1 - t_0)) \| + \| x \|)$ is bounded. Thus, the final limit is zero because $\lim_{t_0 \to 1}(1 - t_0) = 0$, and it follows that θ is continuous.

It is evident from the construction that $\partial_0(\theta) = \phi$. Since this procedure may be applied to any singular k-simplex, $k \geq 0$, there is a unique extension to a homomorphism

$$T: \quad S_k(X) \to S_{k+1}(X)$$

such that $\partial_0 \circ T = $ identity. More generally we have for ϕ a singular k-simplex,

$$\partial_i(T(\phi))(t_0, \ldots, t_k) = T(\phi)(t_0, \ldots, t_{i-1}, 0, t_i, \ldots, t_k)$$

$$= (1 - t_0)\left(\phi\left(\frac{t_1}{1 - t_0}, \ldots, \frac{t_{i-1}}{1 - t_0}, 0, \frac{t_i}{1 - t_0}, \ldots, \frac{t_k}{1 - t_0}\right)\right) + t_0 x.$$

On the other hand,

$$T(\partial_{i-1}(\phi))(t_0, \ldots, t_k) = (1 - t_0)\left(\partial_{i-1}\phi\left(\frac{t_1}{1 - t_0}, \ldots, \frac{t_k}{1 - t_0}\right) + t_0 x\right)$$

$$= (1 - t_0) \cdot \phi\left(\frac{t_1}{1 - t_0}, \ldots, \frac{t_{i-1}}{1 - t_0}, 0, \frac{t_i}{1 - t_0}, \ldots, \frac{t_k}{1 - t_0}\right) + t_0 x.$$

Thus, for $1 \le i \le k + 1$,

$$\partial_i T\phi = T(\partial_{i-1}\phi).$$

Now let ϕ be any singular k-simplex

$$\partial T\phi = \partial_0 T\phi + \sum_{i=1}^{k+1} (-1)^i \partial_i T(\phi)$$

$$= \partial_0 T\phi + \sum_{i=1}^{k+1} (-1)^i \partial_i T(\phi) - \left[\sum_{i=1}^{k+1} (-1)^i T\partial_{i-1}(\phi) + \sum_{j=0}^{k} (-1)^j T\partial_j\phi \right]$$

$$= \phi - T\partial\phi.$$

So we have constructed a homomorphism $T: S_k(X) \to S_{k+1}(X)$ with the property that $\partial T + T\partial$ is the identity homomorphism on $S_k(X)$, whenever $k \ge 1$.

Now let z be an element of $Z_p(X)$. From the above, for $p > 0$, $(\partial T + T\partial)z = z$. Now since z is a cycle, $T\partial z = 0$. Thus, $z = \partial(Tz)$ and z is in $B_p(X)$. This implies that $H_p(X) = 0$ for all $p > 0$. $\square$

The construction used in proving Theorem 1.8 is a special case of a *chain homotopy* between chain complexes. Suppose $C = \{C_i, \partial\}$ and $C' = \{C_i', \partial'\}$ are chain complexes and

$$T: \quad C \to C'$$

is a homomorphism of graded groups of degree one (but not necessarily a chain map). Then consider the homomorphism

$$\partial' T + T\partial: \quad C \to C'$$

of degree zero. This will be a chain map because

$$\partial'(\partial' T + T\partial) = \partial'\partial' T + \partial' T\partial = \partial' T\partial = \partial' T\partial + T\partial\partial = (\partial' T + T\partial)\partial.$$

This chain map $(\partial' T + T\partial)$ induces a homomorphism on homology

$$(\partial' T + T\partial)_*: \quad H_p(C) \to H_p(C') \qquad \text{for each} \quad p.$$

Now if $z \in Z_p(C)$,

$$(\partial' T + T\partial)(z) = \partial' T(z)$$

which is in $B_p(C')$. Thus, $(\partial' T + T\partial)_*$ is the zero homomorphism for each p.

Given chain maps f and $g\colon C \to C'$, f and g are *chain homotopic* if there exists a homomorphism $T\colon C \to C'$ of degree one with $\partial' T + T\partial = f - g$.

1.9   PROPOSITION   If f and $g\colon C \to C'$ are chain homotopic chain maps, then $f_* = g_*$ as homomorphisms from $H_*(C)$ to $H_*(C')$.

PROOF   This follows immediately since if $T\colon C \to C'$ is a chain homotopy between f and g, then

$$0 = (\partial' T + T\partial)_* = (f - g)_* = f_* - g_*. \qquad \square$$

As a special case, suppose that f and $g\colon X \to Y$ are maps for which the induced chain maps

$$f_{\#} \text{ and } g_{\#}\colon \quad S_*(X) \to S_*(Y)$$

are chain homotopic. If T is a chain homotopy between $f_{\#}$ and $g_{\#}$, then T may be interpreted geometrically in the following way.

Let ϕ be a singular n-simplex in X. Then $T(\phi)$ may be viewed as a continuous deformation of $f_{\#}(\phi)$ into $g_{\#}(\phi)$. From Figure 1.4, $T(\phi)$ appears as a prism with ends $f_{\#}(\phi)$ and $g_{\#}(\phi)$ and sides $T(\partial\phi)$. Thus, it is reasonable that

$$\partial T(\phi) = f_{\#}(\phi) - g_{\#}(\phi) - T(\partial\phi),$$

which is the algebraic requirement for T to be a chain homotopy.

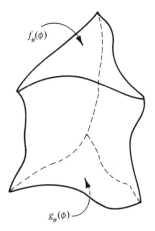

Figure 1.4

If the chain $c = \sum m_i \phi_i$ is an n-cycle in X, then $f_\#(c)$ and $g_\#(c)$ are n-cycles in Y. $T(c)$ is a collection of integral multiples of such prisms and the algebraic sum of the sides must be zero since $\partial c = 0$. Thus, the boundary of $T(c)$ is the algebraic sum of the ends of the prisms, which is $f_\#(c) - g_\#(c)$, so that $f_\#(c)$ and $g_\#(c)$ are homologous cycles in Y.

Given spaces X and Y, two maps $f_0, f_1 : X \to Y$ are *homotopic* if there exists a map

$$F: \ X \times I \to Y, \qquad I = [0, 1],$$

with $F(x, 0) = f_0(x)$ and $F(x, 1) = f_1(x)$, for all x in X. The map F is a *homotopy* between f_0 and f_1. Equivalently a homotopy is a family of maps $\{f_t\}_{0 \le t \le 1}$ from X to Y varying continuously with t. It is evident that the homotopy relation is an equivalence relation on the set of all maps from X to Y. It is customary to denote by $[X, Y]$ the set of homotopy classes of maps.

1.10   THEOREM   If $f_0, f_1 : X \to Y$ are homotopic maps, then $f_{0*} = f_{1*}$ as homomorphisms from $H_*(X)$ to $H_*(Y)$.

PROOF   The idea of the proof is quite simple: if z is a cycle in X, then the images of z under f_0 and f_1 will be cycles in Y. Since f_0 may be continuously deformed into f_1, the image of z under f_0 should admit a similar continuous deformation into the image of z under f_1. This should imply that the two images are homologous cycles. We now proceed to put these geometric ideas into the current algebraic framework.

In view of Proposition 1.9 it will be sufficient to show that the chain maps $f_{0\#}, f_{1\#} : S_*(X) \to S_*(Y)$ are chain homotopic. Let

$$F: \ X \times I \to Y$$

be a homotopy between f_0 and f_1. Define maps

$$g_0, g_1 : \ X \to X \times I$$

by $g_0(x) = (x, 0)$ and $g_1(x) = (x, 1)$:

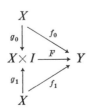

Then in the diagram each triangle is commutative, that is, $f_0 = F \circ g_0$ and $f_1 = F \circ g_1$.

Now suppose that $g_{0\#}$ and $g_{1\#}$ are chain homotopic as chain maps from $S_*(X)$ to $S_*(X \times I)$. This would mean that there exists a homomorphism

$$T: \quad S_*(X) \rightarrow S_*(X \times I)$$

of degree one with $\partial T + T\partial = g_{0\#} - g_{1\#}$. Applying $F_\#$ to both sides gives

$$F_\#(\partial T + T\partial) = F_\#(g_{0\#} - g_{1\#}) \qquad \text{or} \qquad \partial(F_\# T) + (F_\# T)\partial = f_{0\#} - f_{1\#}.$$

Then $F_\# T$ is a homomorphism from $S_*(X)$ to $S_*(Y)$ of degree one and is a chain homotopy between $f_{0\#}$ and $f_{1\#}$. Therefore, it is sufficient to show that $g_{0\#}$ and $g_{1\#}$ are chain homotopic.

For σ_n the standard n-simplex denote by $\tau_n \in S_n(\sigma_n)$ the element represented by the identity map. Note that if $\phi: \sigma_n \rightarrow X$ is any singular n-simplex in X, then the induced homomorphism

$$\phi_\#: \quad S_n(\sigma_n) \rightarrow S_n(X)$$

has $\phi_\#(\tau_n) = \phi$. It is evident that every singular n-simplex in X can be exhibited as the image of τ_n in this manner. Our technique of proof then will be to first give a construction involving τ_n and then extend it to all of $S_n(X)$ by the above approach.

We construct a chain homotopy T between $g_{0\#}$ and $g_{1\#}$ inductively on the dimension of the chain group. To do the inductive step first, suppose that $n > 0$ and for all spaces X and integers $i < n$ there is a homomorphism

$$T: \quad S_i(X) \rightarrow S_{i+1}(X \times I)$$

such that $\partial T + T\partial = g_{0\#} - g_{1\#}$. Assume further that this is natural in the sense that given any map $h: X \rightarrow W$ of spaces, commutativity holds in the diagram

$$
\begin{array}{ccc}
S_i(X) & \xrightarrow{\;T_X\;} & S_{i+1}(X \times I) \\
\downarrow{\scriptstyle h_\#} & & \downarrow{\scriptstyle (h \times \mathrm{id})_\#} \\
S_i(W) & \xrightarrow{\;T_W\;} & S_{i+1}(W \times I)
\end{array}
$$

for all $i < n$.

To define T on the n-chains of X, it is sufficient to define T on the singular n-simplices. So let $\phi: \sigma_n \rightarrow X$ be a singular n-simplex and recall that $\phi_\#(\tau_n) = \phi$. Thus, by defining $T_{\sigma_n}: S_n(\sigma_n) \rightarrow S_{n+1}(\sigma_n \times I)$, the naturality of the construction will require that

$$T_X(\phi) = T_X(\phi_\#(\tau_n)) = (\phi \times \mathrm{id})_\#(T_{\sigma_n}(\tau_n)).$$

So to define T_X it is sufficient to define T_{σ_n} on $S_n(\sigma_n)$.

Let d be a singular n-simplex in σ_n and consider the chain in $S_n(\sigma_n \times I)$ given by

$$c = g_{0\#}(d) - g_{1\#}(d) - T_{\sigma_n}(\partial d),$$

which is defined by the induction hypothesis since ∂d is in $S_{n-1}(\sigma_n)$. Note that from the preceding discussion, c corresponds to the boundary of a certain prism in σ_n. Then

$$\begin{aligned}
\partial c &= \partial g_{0\#}(d) - \partial g_{1\#}(d) - \partial T_{\sigma_n}(\partial d) \\
&= g_{0\#}(\partial d) - g_{1\#}(\partial d) - [g_{0\#}(\partial d) - g_{1\#}(\partial d) - T_{\sigma_n}\partial(\partial d)] \\
&= 0.
\end{aligned}$$

Thus, c is a cycle of dimension n in the convex set $\sigma_n \times I$. From Theorem 1.8 it follows that c is also a boundary. So let $b \in S_{n+1}(\sigma_n \times I)$ with $\partial b = c$. Geometrically b is the solid prism of which c is the boundary. Then define

$$T_{\sigma_n}(d) = b$$

and observe that

$$\partial T(d) + T\partial(d) = g_{0\#}(d) - g_{1\#}(d).$$

Now for any singular n-simplex $\phi: \sigma_n \to X$ define, as before,

$$T_X(\phi) = (\phi \times \mathrm{id})_\# T_{\sigma_n}(\tau_n).$$

So defined on the generators there is a unique extension to a homomorphism

$$T_X: \quad S_n(X) \to S_{n+1}(X \times I).$$

This inductive construction indicates the proper definition for T on 0-chains. Recall that σ_0 is a point and consider the chain c in $S_0(\sigma_0 \times I)$ given by

$$c = g_{0\#}(\tau_0) - g_{1\#}(\tau_0).$$

Take a singular 1-simplex b in $\sigma_0 \times I$ with boundary $g_{0\#}(\tau_0) - g_{1\#}(\tau_0)$ and define $T_{\sigma_0}(\tau_0) = b$. This defines T on 0-chains by the same technique.

Finally it must be noted that in the definition given for T_X on n-chains of X,

$$\partial T_X + T_X \partial = g_{0\#} - g_{1\#}$$

and that the construction is suitably natural with respect to maps $h: X \to W$. Note that if ϕ is a singular n-simplex in X,

$$g_{0\#}(\phi) = g_{0\#}\phi_{\#}(\tau_n) = (\phi \times \mathrm{id})_{\#}g_{0\#}(\tau_n)$$

and similarly

$$g_{1\#}(\phi) = g_{1\#}\phi_{\#}(\tau_n) = (\phi \times \mathrm{id})_{\#}g_{1\#}(\tau_n).$$

Now consider

$$
\begin{aligned}
\partial T(\phi) + T\partial(\phi) &= \partial T\phi_{\#}(\tau_n) + T\partial\phi_{\#}(\tau_n) \\
&= \partial(\phi \times \mathrm{id})_{\#}T(\tau_n) + T\phi_{\#}\partial(\tau_n) \\
&= (\phi \times \mathrm{id})_{\#}\partial T(\tau_n) + (\phi \times \mathrm{id})_{\#}T\partial(\tau_n) \\
&= (\phi \times \mathrm{id})_{\#}(g_{0\#}(\tau_n) - g_{1\#}(\tau_n)) \\
&= g_{0\#}(\phi) - g_{1\#}(\phi).
\end{aligned}
$$

The naturality follows similarly.

Therefore, T_X gives a chain homotopy between $g_{0\#}$ and $g_{1\#}$, and we have completed the proof that $f_{0*} = f_{1*}$. $\square$

Note that this generalizes the approach in Theorem 1.8. There we used the fact that, since X was convex, the identity map was homotopic to the map sending all of X into the point x. Thus in positive dimensions the identity homomorphism and the trivial homomorphism agree, and the positive dimensional homology of X is trivial.

Let $f: X \to Y$ and $g: Y \to X$ be maps of topological spaces. If the compositions $f \circ g$ and $g \circ f$ are each homotopic to the respective identity map, then f and g are *homotopy inverses* of each other. A map $f: X \to Y$ is a *homotopy equivalence* if f has a homotopy inverse; in this case X and Y are said to have the *same homotopy type*.

1.11 PROPOSITION If $f: X \to Y$ is a homotopy equivalence, then $f_*: H_n(X) \to H_n(Y)$ is an isomorphism for each n.

PROOF If g is a homotopy inverse for f, then by Theorem 1.10 $f_* \circ g_* = (f \circ g)_* = $ identity and $g_* \circ f_* = (g \circ f)_* = $ identity so that $g_* = f_*^{-1}$ and f_* is an isomorphism. $\square$

Suppose that $i: A \to X$ is the inclusion map of a subspace A of X. A map $g: X \to A$ such that $g \circ i$ is the identity on A is a *retraction* of X onto A. If furthermore the composition $i \circ g: X \to X$ is homotopic to the identity, then g is a *deformation retraction* and A is a *deformation retract* of X. Note that in this case the inclusion i is a homotopy equivalence.

1.12 COROLLARY If $i: A \to X$ is the inclusion of a retract A of X, then $i_*: H_*(A) \to H_*(X)$ is a monomorphism onto a direct summand. If A is a deformation retract of X, then i_* is an isomorphism.

PROOF The second statement follows immediately from Proposition 1.11. To prove the first, let $g: X \to A$ be a retraction. Then

$$g_* \circ i_* = (g \circ i)_* = (\mathrm{id})_* = \text{identity on} H_*(A).$$

Hence, i_* is a monomorphism.

Define subgroups of $H_*(X)$ by $G_1 = \text{image } i_*$ and $G_2 = \text{kernel } g_*$. Let $\alpha \in G_1 \cap G_2$, so that $\alpha = i_*(\beta)$ for some $\beta \in H_*(A)$ and $g_*(\alpha) = 0$. However

$$0 = g_*(\alpha) = g_* i_*(\beta) = \beta$$

so that $\alpha = i_*(\beta)$ must be zero. On the other hand, let $\gamma \in H_*(X)$. Then

$$\gamma = i_* g_*(\gamma) + (\gamma - i_* g_*(\gamma))$$

expresses γ as the sum of an element in G_1 and an element in G_2. Therefore, $H_*(X) \approx G_1 \oplus G_2$ and the proof is complete. $\square$

A triple $C \xrightarrow{f} D \xrightarrow{g} E$ of abelian groups and homomorphisms is *exact* if image $f = $ kernel g. A sequence of abelian groups and homomorphisms

$$\cdots \to G_1 \xrightarrow{f_1} G_2 \xrightarrow{f_2} G_3 \xrightarrow{f_3} \cdots \xrightarrow{f_{n-1}} G_n \xrightarrow{f_n} \cdots$$

is exact if each triple is exact. An exact sequence

$$0 \to C \xrightarrow{f} D \xrightarrow{g} E \to 0$$

is called *short exact*. This is a generalization of the concept of isomorphism in the sense that $h: G_1 \to G_2$ is an isomorphism if and only if

$$0 \to G_1 \xrightarrow{h} G_2 \to 0$$

is exact.

Note that in a short exact sequence as above, f is a monomorphism and identifies C with a subgroup $C' \subseteq D$. Also g is an epimorphism with kernel C'. Thus up to isomorphism the sequence is just

$$0 \to C' \xrightarrow{i} D \xrightarrow{\pi} D/C' \to 0.$$

Suppose now that $C = \{C_n\}$, $D = \{D_n\}$ and $E = \{E_n\}$ are chain complexes and

$$0 \to C \xrightarrow{f} D \xrightarrow{g} E \to 0$$

is a short exact sequence where f and g are chain maps of degree zero. Hence, for each p there is an associated triple of homology groups,

$$H_p(C) \xrightarrow{f_*} H_p(D) \xrightarrow{g_*} H_p(E).$$

We now want to examine precisely how this deviates from being short exact.

So we are assuming that we have an infinite diagram in which the rows are short exact sequences and each square is commutative.

$$
\begin{array}{ccccccccc}
 & \vdots & & \vdots & & \vdots & & & \\
 & \downarrow & & \downarrow & & \downarrow & & & \\
0 \to & C_n & \xrightarrow{f} & D_n & \xrightarrow{g} & E_n & \longrightarrow & 0 \\
 & \downarrow{\scriptstyle\partial} & & \downarrow{\scriptstyle\partial} & & \downarrow{\scriptstyle\partial} & & & \\
0 \to & C_{n-1} & \xrightarrow{f} & D_{n-1} & \xrightarrow{g} & D_{n-1} & \longrightarrow & 0 \\
 & \downarrow{\scriptstyle\partial} & & \downarrow{\scriptstyle\partial} & & \downarrow{\scriptstyle\partial} & & & \\
 & \vdots & & \vdots & & \vdots & & &
\end{array}
$$

Let $z \in Z_n(E)$, that is, $z \in E_n$ and $\partial z = 0$. Since g is an epimorphism, there exists an element $d \in D_n$ with $g(d) = z$. From the fact that g is a chain map we have

$$g(\partial d) = \partial(g(d)) = \partial z = 0.$$

The exactness implies that ∂d is in the image of f, so let $c \in C_{n-1}$ with $f(c) = \partial d$. Note that

$$f(\partial c) = \partial f(c) = \partial(\partial d) = 0,$$

and since f is a monomorphism, ∂c must be zero, and $c \in Z_{n-1}(C)$.

The correspondence $z \to c$ of $Z_n(E)$ into $Z_{n-1}(C)$ is not a well-defined function from cycles to cycles due to the number of possible choices in the construction. However, we now show that the associated correspondence on the homology groups is a well-defined homomorphism.

Let $z, z' \in Z_n(E)$ be homologous cycles. So there exists an element $e \in E_{n+1}$ with $\partial(e) = z - z'$. Let $d, d' \in D_n$ with $g(d) = z$, $g(d') = z'$, and $c, c' \in C_{n-1}$ with $f(c) = \partial d$, $f(c') = \partial d'$. We must show that c and c' are homologous cycles.

There exists an element $a \in D_{n+1}$ with $g(a) = e$. By the commutativity

$$g(\partial a) = \partial g(a) = \partial e = z - z',$$

so we observe that $(d - d') - \partial a$ is in the kernel of g, hence also in the image of f. Let $b \in C_n$ with $f(b) = (d - d') - \partial a$. Now we have

$$f(\partial b) = \partial f(b) = \partial(d - d' - \partial a) = \partial d - \partial d'$$
$$= f(c) - f(c') = f(c - c').$$

Since f is one to one, it follows that $c - c' = \partial b$ and c and c' are homologous cycles. Therefore, the correspondence induced on the homology groups is well defined and obviously must be a homomorphism.

This homomorphism is denoted by $\Delta : H_n(E) \to H_{n-1}(C)$ and called the *connecting homomorphism* for the short exact sequence

$$0 \to C \xrightarrow{f} D \xrightarrow{g} E \to 0.$$

1.13   THEOREM   If $0 \to C \xrightarrow{f} D \xrightarrow{g} E \to 0$ is a short exact sequence of chain complexes and degree zero chain maps, then the long exact sequence

$$\cdots \xrightarrow{f_*} H_n(D) \xrightarrow{g_*} H_n(E) \xrightarrow{\Delta} H_{n-1}(C) \xrightarrow{f_*} H_{n-1}(D) \xrightarrow{g_*} \cdots$$

is exact.

Exercise 4   Prove Theorem 1.13.   □

It is important to note that the construction of the connecting homomorphism is suitably natural. That is, if

$$
\begin{array}{ccccccccc}
0 & \to & C & \xrightarrow{f} & D & \xrightarrow{g} & E & \to & 0 \\
 & & \downarrow{\alpha} & & \downarrow{\beta} & & \downarrow{\gamma} & & \\
0 & \to & C' & \xrightarrow{f'} & D' & \xrightarrow{g'} & E' & \to & 0
\end{array}
$$

is a diagram of chain complexes and degree zero chain maps in which the rows are exact and the rectangles are commutative, then commutativity holds in each rectangle of the associated diagram

$$\cdots \to H_n(D) \xrightarrow{g_*} H_n(E) \xrightarrow{\Delta} H_{n-1}(C) \xrightarrow{f_*} H_{n-1}(D) \to \cdots$$

$$\Big\downarrow \beta_* \qquad\qquad \Big\downarrow \gamma_* \qquad\qquad \Big\downarrow \alpha_* \qquad\qquad \Big\downarrow \beta_*$$

$$\cdots \to H_n(D') \xrightarrow{g_*'} H_n(E') \xrightarrow{\Delta'} H_{n-1}(C') \xrightarrow{f_*'} H_{n-1}(D') \to \cdots$$

Let X be a topological space and $A \subseteq X$ a subspace. The *interior* of A (Int A) is the union of all open subsets of X which are contained in A, or equivalently the maximal subset of A which is open in X. A collection $\mathcal{U}$ of subsets of X is a covering of X if $X \subseteq \bigcup_{u \in \mathcal{U}} U$. Given a collection $\mathcal{U}$, let int $\mathcal{U}$ be the collection of interiors of elements of $\mathcal{U}$. We will be interested in those $\mathcal{U}$ for which int $\mathcal{U}$ is a covering of X.

For $\mathcal{U}$ any covering of X, denote by $S_n{}^{\mathcal{U}}(X)$ the subgroup of $S_n(X)$ generated by the singular n-simplices $\phi: \sigma_n \to X$ for which $\phi(\sigma_n)$ is contained in some $U \in \mathcal{U}$. Then for each i

$$\text{image } \partial_i \phi \subseteq \text{image } \phi$$

so that the total boundary

$$\partial: \ S_n{}^{\mathcal{U}}(X) \to S_{n-1}^{\mathcal{U}}(X).$$

So associated with any covering $\mathcal{U}$ of X there is a chain complex $S_*^{\mathcal{U}}(X)$ and the natural inclusion

$$i: \ S_*^{\mathcal{U}}(X) \to S_*(X)$$

is a chain map. Note that if $\mathcal{V}$ is a covering of a space Y and $f: X \to Y$ is a map such that for each $U \in \mathcal{U}$, $f(U)$ is contained in some V of $\mathcal{V}$, then there is a chain map

$$f_{\#}: \ S_*^{\mathcal{U}}(X) \to S_*^{\mathcal{V}}(Y)$$

and $f_{\#} \circ i_X = i_Y \circ f_{\#}$.

We are now ready for the theorem which will serve as the essential computational tool in studying the homology groups of spaces.

1.14   THEOREM If $\mathcal{U}$ is a family of subsets of X such that Int $\mathcal{U}$ is a covering of X, then

$$i_*: \ H_n(S_*^{\mathcal{U}}(X)) \to H_n(X)$$

is an isomorphism for each n.

PROOF   See Appendix I.   □

The proof is deferred to an appendix to avoid a lengthy interruption of the exposition. It should not be assumed that this implies the proof is either irrelevant or uninteresting. Indeed this argument characterizes the basic difference between homology theory and homotopy theory. Intuitively the approach to proving this theorem is evident. Given a chain c in X we must construct a chain c' in X such that c' is in the image of i and $\partial c = \partial c'$. Moreover, if c is a cycle we will want c' to be homologous to c. This is done by "subdividing" the chain c repeatedly until the resulting chain is the desired c'. The technique of subdivision is possible in homology theory because an n-simplex may be subdivided into a collection of smaller n-simplices. However, the subdivision of a sphere does not result in a collection of smaller spheres. It is the absence of such a construction, that makes the computation of homotopy groups extremely difficult for spaces as simple as a sphere.

To see the requirement that Int $\mathcal{U}$ covers X is essential, let $X = S^1$, $x_0 \in S^1$, and $\mathcal{U} = \{\{x_0\}, S^1 - \{x_0\}\}$. Then any chain c in $S_1^{\mathcal{U}}(S^1)$ may be uniquely written as the sum of a chain c_1 in $\{x_0\}$ and a chain c_2 in $S^1 - \{x_0\}$. Moreover, since the image of c_2 is contained in a compact subset of $S^1 - \{x_0\}$, c will be a cycle if and only if each of c_1 and c_2 are cycles. Now both c_1 and c_2 must then also be boundaries; hence, $H_1(S_*^{\mathcal{U}}(S^1)) = 0$. However, it will soon be shown that $H_1(S^1) \approx Z$.

The first application of Theorem 1.14 will be the development of a technique for studying the homology of a space X in terms of the homology of the components of a covering $\mathcal{U}$ of X. In the simplest nontrivial case the covering $\mathcal{U}$ consists of two subsets U and V for which Int $U \cup$ Int $V = X$. For convenience let A' be the set of all singular n-simplices in U and A'' be the set of all singular n-simplices in V. Then

$$S_n(U) = F(A'), \qquad\qquad S_n(V) = F(A''),$$
$$S_n(U \cap V) = F(A' \cap A''), \qquad S_n^{\mathcal{U}}(X) = F(A' \cup A'').$$

Note that there is a natural homomorphism

$$h: \quad F(A') \oplus F(A'') \to F(A' \cup A'')$$

given by

$$h(a_i', a_j'') = a_i' + a_j''.$$

It is not difficult to see that h is an epimorphism. On the other hand, there is the homomorphism

$$g: \quad F(A' \cap A'') \to F(A') \oplus F(A'')$$

given by

$$g(b_i) = (b_i, -b_i).$$

It follows immediately that g is a monomorphism and $h \circ g = 0$. Now suppose

$$h(\sum n_i a_i', \ \sum m_j a_j'') = 0.$$

That is

$$\sum n_i a_i' + \sum m_j a_j'' = 0.$$

Since these are free abelian groups, the only way this can happen is for each nonzero n_i, $a_i' = a_j''$ for some j and furthermore $m_j = -n_i$. All non-zero coefficients m_j must appear in this manner. This implies that all a_i' are in $A' \cap A''$ and if $x = \sum n_i a_i'$, then $\sum m_j a_j'' = -x$. Hence,

$$x \in F(A' \cap A'') \qquad \text{and} \qquad g(x) = (\sum n_i a_i', \ \sum m_j a_j'').$$

This proves that the kernel of h is contained in the image of g, and interpreting these facts in terms of the chain groups gives for each n a short exact sequence

$$0 \to S_n(U \cap V) \xrightarrow{g_{\#}} S_n(U) \oplus S_n(V) \xrightarrow{h_{\#}} S_n^{\mathfrak{U}}(X) \to 0.$$

Define a chain complex $S_*(U) \oplus S_*(V)$ by setting $(S_*(U) \oplus S_*(V))_n = S_n(U) \oplus S_n(V)$ and letting the boundary operator be the usual boundary on each component. Then the above sequence becomes a short exact sequence of chain complexes and degree zero chain maps.

By Theorem 1.13 there is associated a long exact sequence of homology groups,

$$\cdots \xrightarrow{\Delta} H_n(U \cap V) \xrightarrow{g_*} H_n(S_*(U) \oplus S_*(V)) \xrightarrow{h_*} H_n(S_*^{\mathfrak{U}}(X))$$
$$\xrightarrow{\Delta} H_{n-1}(U \cap V) \to \cdots.$$

From the definition of the chain complex it is evident that $H_n(S_*(U) \oplus S_*(V)) \approx H_n(U) \oplus H_n(V)$, and by Theorem 1.14 we have

$$H_n(S_*^{\mathfrak{U}}(X)) \approx H_n(X).$$

Incorporating these isomorphisms into the long exact sequence, we have established the *Mayer–Vietoris sequence*

$$\cdots \xrightarrow{\;\Delta\;} H_n(U \cap V) \xrightarrow{\;g_*\;} H_n(U) \oplus H_n(V) \xrightarrow{\;h_*\;} H_n(X)$$
$$\xrightarrow{\;\Delta\;} H_{n-1}(U \cap V) \to \cdots.$$

Note that if we define by

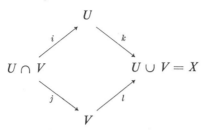

the respective inclusion maps, then $g_*(x) = (i_*(x), -j_*(x))$ and $h_*(y, z) = k_*(y) + l_*(z)$. The connecting homomorphism Δ may be interpreted geometrically as follows: any homology class ω in $H_n(X)$ may be represented by a cycle $c + d$ where c is a chain in U and d is a chain in V. (This follows from Theorem 1.14.) Then $\Delta(\omega)$ is represented by the cycle ∂c in $U \cap V$.

The construction of the Mayer–Vietoris sequence is natural in the sense that if X' is a space, U' and V' are subsets with Int $U' \cup$ Int $V' = X'$, and $f: X \to X'$ is a map for which $f(U) \subseteq U'$ and $f(V) \subseteq V'$, then commutativity holds in each rectangle of the diagram

$$
\begin{array}{ccccccc}
\cdots \xrightarrow{\Delta} & H_n(U \cap V) & \xrightarrow{g_*} & H_n(U) \oplus H_n(V) & \xrightarrow{h_*} & H_n(X) & \xrightarrow{\Delta} H_{n-1}(U \cap V) \to \cdots \\
& \downarrow{f_*} & & \downarrow{f_* \oplus f_*} & & \downarrow{f_*} & \qquad \downarrow{f_*} \\
\cdots \xrightarrow{\Delta'} & H_n(U' \cap V') & \xrightarrow{g_*'} & H_n(U') \oplus H_n(V') & \xrightarrow{h_*'} & H_n(X') & \xrightarrow{\Delta'} H_{n-1}(U' \cap V') \to \cdots
\end{array}
$$

Example Let $X = S^1$ and denote by z and z' the north and south poles, respectively, and by x and y the points on the equator (Figure 1.5). Let $U = S^1 - \{z'\}$ and $V = S^1 - \{z\}$. Then in the Mayer–Vietoris sequence associated with this covering we have

$$H_1(U) \oplus H_1(V) \xrightarrow{\;h_*\;} H_1(S^1) \xrightarrow{\;\Delta\;} H_0(U \cap V) \xrightarrow{\;g_*\;} H_0(U) \oplus H_0(V).$$

The first term is zero since U and V are contractible. Thus, Δ is a monomorphism and $H_1(S^1)$ will be isomorphic to the image of $\Delta =$ the kernel of g_*. An element of $H_0(U \cap V) \approx Z \oplus Z$ may be written in the form $ax + by$, where a and b are integers.

Now

$$g_*(ax + by) = (i_*(ax + by), -j_*(ax + by)).$$

Since U and V are pathwise connected, $i_*(ax + by) = 0$ if and only if $a = -b$ and similarly for j_*. Thus the kernel of g_* is the subgroup of $H_0(U \cap V)$ consisting of all elements of the form $ax - ay$. This is an infinite cyclic subgroup generated by $x - y$. Therefore, we conclude that

$$H_1(S^1) \approx Z.$$

To give geometrically a generator ω for this group, we must represent ω by the sum of two chains, $c + d$, where c is in U and d is in V, for which $\partial(c) = x - y = -\partial d$. The chains c and d may be chosen as shown in Figure 1.6.

For any integer $n > 1$ the portion of the Mayer–Vietoris sequence

$$H_n(U) \oplus H_n(V) \xrightarrow{h_*} H_n(S^1) \xrightarrow{\Delta} H_{n-1}(U \cap V)$$

has the two end terms equal to zero; hence, $H_n(S^1) = 0$.

This completes the determination of the homology of S^1. We now proceed inductively to compute the homology of S^n for each n. Recall that

$$S^n = \{(x_1, \ldots, x_{n+1}) \mid x_i \in R, \sum x_i^2 = 1\} \subseteq R^{n+1}.$$

In the usual fashion consider $R^n \subseteq R^{n+1}$ as all points of the form $(x_1, \ldots, x_n, 0)$. Under this inclusion $S^{n-1} \subseteq S^n$ as the "equator." Denote by $z = (0, \ldots, 0, 1)$ and $z^1 = (0, \ldots, 0, -1)$ the north and south poles of S^n. Then by stereographic projection $S^n - \{z\}$ is homeomorphic to R^n, and similarly for $S^n - \{z^1\}$. Furthermore, $S^n - \{z \cup z^1\}$ is homeomorphic to $R^n - \{\text{origin}\}$.

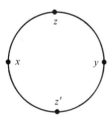

Figure 1.5

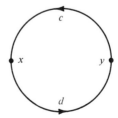

Figure 1.6

Exercise 5 Show that S^{n-1} is a deformation retract of $\mathbb{R}^n - \{\text{origin}\}$.

Now let $U = S^n - \{z\}$, $V = S^n - \{z^1\}$ so that $U \cap V = S^n - \{z \cup z^1\}$. Then by the observations and the exercise above, the Mayer–Vietoris sequence for this covering becomes

$$H_m(\mathbb{R}^n) \oplus H_m(\mathbb{R}^n) \xrightarrow{h_*} H_m(S^n) \xrightarrow{\Delta} H_{m-1}(S^{n-1}) \xrightarrow{g_*} H_{m-1}(\mathbb{R}^n) \oplus H_{m-1}(\mathbb{R}^n).$$

For $m > 1$, the end terms are zero so that Δ is an isomorphism. For $m = 1$ and $n > 1$, g_* and Δ must both be monomorphisms so that $H_1(S^n) = 0$. This furnishes the inductive step in the proof of the following:

1.15 THEOREM For any integer $n \geq 0$, $H_*(S^n)$ is a free abelian group with two generators, one in dimension zero and one in dimension n. $\square$

1.16 COROLLARY For $n \neq m$, S^n and S^m do not have the same homotopy type. $\square$

Exercise 6 Using only the tools that we have developed, compute the homology of a two-sphere with two handles (Figure 1.7).

Define the *n-disk* in $\mathbb{R}^n$ to be

$$D^n = \{(x_1, \ldots, x_n) \in \mathbb{R}^n \mid \sum x_i^2 \leq 1\}$$

and note that $S^{n-1} \subseteq D^n$ is its boundary.

1.17 COROLLARY There is no retraction of D^n onto S^{n-1}.

PROOF For $n = 1$ this is obvious since D^1 is connected and S^0 is not. Suppose $n > 1$ and $f: D^n \to S^{n-1}$ is a map such that $f \circ i = $ identity, where i is the inclusion of S^{n-1} in D^n.

This implies that the following diagram of homology groups and induced

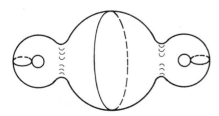

Figure 1.7

homomorphisms is commutative:

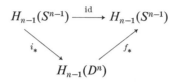

$$H_{n-1}(S^{n-1}) \xrightarrow{\text{id}} H_{n-1}(S^{n-1})$$

However, this gives a factorization of the identity on an infinite cyclic group through zero which is impossible. Therefore, no such retraction f exists.   □

1.18 COROLLARY   (*Brouwer fixed-point theorem*)   Given a map $f: D^n \to D^n$, there exists an x in D^n with $f(x) = x$.

PROOF   Suppose $f: D^n \to D^n$ without fixed points. Define a function $g: D^n \to S^{n-1}$ as follows: for $x \in D^n$ there is a well-defined ray starting at $f(x)$ and passing through x. Define $g(x)$ to be the point at which this ray intersects S^{n-1} (Figure 1.8). Then $g: D^n \to S^{n-1}$ is continuous and $g(x) = x$ for all x in S^{n-1}. But the existence of such a map g contradicts Corollary 1.17. Therefore, f must have a fixed point.   □

Exercise 7   Show that Corollary 1.18 implies Corollary 1.17.

Let $n \geq 1$ and suppose that $f: S^n \to S^n$ is a map. Choose a generator α of $H_n(S^n) \approx Z$ and note that the homomorphism induced by f on $H_n(S^n)$ has $f_*(\alpha) = m \cdot \alpha$ for some integer m. This integer is independent of the choice of the generator since $f_*(-\alpha) = -f_*(\alpha) = -m \cdot \alpha = m \cdot (-\alpha)$. The integer m is the *degree* of f, denoted $d(f)$. This is often referred to as the *Brouwer degree* as a result of his efforts in developing this idea. The degree of a map is a direct generalization of the "winding number" associated with a map from the circle into the nonzero complex numbers.

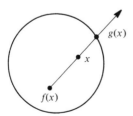

Figure 1.8

The following basic properties of the degree of a map are immediate consequences of our previous results:

(a) $d(\text{identity}) = 1$;
(b) if f and $g: S^n \to S^n$ are maps, $d(f \circ g) = d(f) \cdot d(g)$;
(c) $d(\text{constant map}) = 0$;
(d) if f and g are homotopic, then $d(f) = d(g)$;
(e) if f is a homotopy equivalence then $d(f) = \pm 1$.

A slightly less obvious property (a future exercise) is that there exist maps of any integral degree on S^n whenever $n > 0$. All these properties are results of homology theory, and as such are easily obtained. A much more sophisticated property is the homotopy theoretic result of Hopf, which is the converse of property (d), if $d(f) = d(g)$ then f and g are homotopic. Thus, the degree is a complete algebraic invariant for studying homotopy classes of maps from S^n to S^n.

1.19 PROPOSITION   Let $n > 0$ and define $f: S^n \to S^n$ by

$$f(x_1, x_2, \ldots, x_{n+1}) = (-x_1, x_2, \ldots, x_{n+1}).$$

Then $d(f) = -1$.

PROOF   First consider the case $n = 1$ (Figure 1.9). As before let $z = (0, 1)$, $z^1 = (0, -1)$ and $x = (-1, 0)$, $y = (1, 0)$. The covering $U = S^1 - \{z^1\}$ and $V = S^1 - \{z\}$ has the property that $f(U) \subseteq U$ and $f(V) \subseteq V$.
Thus, by the naturality of the Mayer–Vietoris sequence the diagram

$$
\begin{array}{ccc}
0 \to H_1(S^1) & \overset{\Delta}{\longrightarrow} & H_0(U \cap V) \\
\downarrow{\scriptstyle f_*} & & \downarrow{\scriptstyle f_{3*}} \\
0 \to H_1(S^1) & \overset{\Delta:}{\longrightarrow} & H_0(U \cap V)
\end{array}
$$

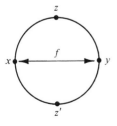

Figure 1.9

has exact rows, and the rectangle commutes where f_3 is the restriction of f. Recall that a generator α of $H_1(S^1)$ was represented by the cycle $c + d$ where $\partial c = x - y = -\partial d$, and $\varDelta(\alpha)$ is represented by $x - y$. Now

$$\varDelta f_*(\alpha) = f_{3*}\varDelta(x) = f_{3*}(x - y) = y - x = -\varDelta(\alpha) = \varDelta(-\alpha).$$

Since $\varDelta$ is a monomorphism, $d(f) = -1$.

Now suppose the conclusion is true in dimension $n - 1 \geq 1$ and consider $S^{n-1} \subseteq S^n$ as before. Taking U and V to be the complements of the south pole and the north pole, respectively, in S^n, the inclusion

$$i: \quad S^{n-1} \to U \cap V$$

is a homotopy equivalence. Since $n \geq 2$, the connecting homomorphism in the Mayer–Vietoris sequence is an isomorphism. Thus, in the diagram

$$
\begin{array}{ccccc}
H_n(S^n) & \xrightarrow[\approx]{\varDelta} & H_{n-1}(U \cap V) & \xleftarrow[\approx]{i_*} & H_{n-1}(S^{n-1}) \\
\downarrow{\scriptstyle f_*} & & \downarrow{\scriptstyle f_{3*}} & & \downarrow{\scriptstyle f_*} \\
H_n(S^n) & \xrightarrow[\approx]{\varDelta} & H_{n-1}(U \cap V) & \xleftarrow[\approx]{i_*} & H_{n-1}(S^{n-1})
\end{array}
$$

each rectangle commutes and the horizontal homomorphisms are isomorphisms. If α is a generator of $H_n(S^n)$,

$$f_*(\alpha) = \varDelta^{-1} f_{3*} \varDelta(\alpha) = \varDelta^{-1} i_* f_* i_*^{-1} \varDelta(\alpha) = -\varDelta^{-1} i_* i_*^{-1} \varDelta(\alpha) = -\alpha.$$

This gives the inductive step and the proof is complete.   □

For a given map $f: S^n \to S^n$, $n \geq 0$, there is associated a map $g: S^{n+1} \to S^{n+1}$ called the *suspension* of f and denoted by Σf. Intuitively, the idea is that the restriction to the equator (S^n) in S^{n+1} should be f and each slice in S^{n+1} parallel to the equator should be mapped into the cor-

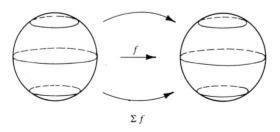

$$\Sigma f$$

Figure 1.10

responding slice in the manner prescribed by f (Figure 1.10). Specifically consider $S^{n+1} \subseteq R^{n+2} = R^{n+1} \times R^1$ so that the points of S^{n+1} are of the form (x, t), where $x \in R^{n+1}$, $t \in R^1$ and $\| x \|^2 + | t |^2 = 1$. Then define

$$\Sigma f(x, t) = \begin{cases} (x, t) & \text{if } x = 0 \\ (\| x \| \cdot f(x/\| x \|), t) & \text{if } x \neq 0. \end{cases}$$

It is not difficult to see that Σf is continuous and has the desired characteristics.

The technique used in proving Proposition 1.19 may be applied to establish the following:

1.20 PROPOSITION If $f: S^n \to S^n$, $n \geq 1$ is a map, then $d(\Sigma f) = d(f)$. $\quad\square$

Note that if $f(x_1, \ldots, x_{n+1}) = (-x_1, \ldots, x_{n+1})$ and $g(x_1, \ldots, x_{n+2}) = (-x_1, \ldots, x_{n+2})$, then $g = \Sigma f$ and Proposition 1.19 is a special case of Proposition 1.20.

1.21 COROLLARY If $f: S^n \to S^n$ is given by

$$f(x_1, \ldots, x_{n+1}) = (x_1, \ldots, -x_i, \ldots, x_{n+1}),$$

then $d(f) = -1$.

PROOF Let $h: S^n \to S^n$ be the map that exchanges the first coordinate and the ith coordinate. Then h is a homeomorphism ($h^{-1} = h$), so $d(h) = \pm 1$. Let $g(x_1, \ldots, x_{n+1}) = (-x_1, \ldots, x_{n+1})$ so that $d(g) = -1$. Then

$$d(f) = d(h \circ g \circ h) = d(h)^2\, d(g) = (\pm 1)^2 (-1) = -1. \quad\square$$

1.22 COROLLARY The antipodal map $A: S^n \to S^n$ defined by $A(x_1, \ldots, x_n) = (-x_1, \ldots, -x_n)$ has $d(A) = (-1)^{n+1}$.

PROOF From Corollary 1.21 A is the composition of $(n + 1)$-maps, all having degree -1. $\quad\square$

Exercise 8 Show that for $n > 0$ and m any integer, there exists a map $f: S^n \to S^n$ of degree m.

1.23 PROPOSITION If $f, g: S^n \to S^n$ are maps with $f(x) \neq g(x)$ for all x in S^n, then g is homotopic to $A \circ f$.

PROOF   Graphically the idea is as follows: since $g(x) \neq f(x)$, the segment in $\mathbb{R}^{n+1}$ from $Af(x)$ to $g(x)$ does not pass through the origin. Thus, projecting out from the origin onto the sphere yields a path in S^n between $Af(x)$ and $g(x)$ (Figure 1.11). These are the paths which produce the desired homotopy. In particular we define a function

$$F: S^n \times I \to S^n$$

by

$$F(x, t) = \frac{(1 - t)Af(x) + t \cdot g(x)}{\| (1 - t)Af(x) + t \cdot g(x) \|}$$

which gives the homotopy explicitly. $\square$

1.24   COROLLARY   If $f: S^{2n} \to S^{2n}$ is a map, then there exists an x in S^{2n} with $f(x) = x$ or there exists a y in S^{2n} with $f(y) = -y$.

PROOF   If $f(x) \neq x$ for all x, then by Proposition 1.23 f is homotopic to A. On the other hand, if $f(x) \neq -x = A(x)$ for all x, then f is homotopic to $A \circ A =$ identity.

When both of these conditions hold, we have

$$d(A) = d(f) = d(\text{identity}).$$

However, $d(A) = (-1)^{2n+1} = -1$ and $d(\text{identity}) = 1$, and the two conditions cannot hold simultaneously. $\square$

1.25   COROLLARY   There is no continuous map $f: S^{2n} \to S^{2n}$ such that x and $f(x)$ are orthogonal for all x. $\square$

Although these ideas have not been defined, S^n is a manifold of dimension n. That is, it is locally homeomorphic to $\mathbb{R}^n$. As such it has a tangent space $T(S^n, x)$ at each point x in S^n. With S^n identified with the unit sphere

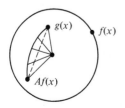

Figure 1.11

in $\mathbb{R}^{n+1}$, $T(S^n, x)$ is the n-dimensional hyperplane in $\mathbb{R}^{n+1}$ which is tangent to S^n at x (Figure 1.12). We may translate this hyperplane to the origin where it becomes the n-dimensional subspace orthogonal to the vector x. Of course, as x varies over S^n, these subspaces will vary accordingly. A *vector field* on S^n is a continuous function assigning to each x in S^n a vector in the corresponding linear subspace. A vector field ϕ is nonzero if $\phi(x) \neq 0$ for each x in S^n.

1.26   COROLLARY   There exists no nonzero vector field on S^{2n}.

PROOF   If ϕ is a nonzero vector field on S^{2n}, then $\psi(x) = \phi(x)/\| \phi(x) \|$ is a vector field on S^{2n} of unit length. Thus, $\psi: S^{2n} \to S^{2n}$ is a map for which $\psi(x)$ is orthogonal to x for each x. But this is impossible by Corollary 1.25. Hence, no such vector field exists. $\square$

Nonzero vector fields always exist on odd-dimensional spheres. A collection of vector fields $\phi_1, \ldots, \phi_k$ on S^n is linearly independent if for each x in S^n the vectors $\phi_1(x), \ldots, \phi_k(x)$ are linearly independent. A famous problem in mathematics is the determination of the maximum number of linearly independent vector fields which exist on S^{2n+1} for each value of n. The work of Hurwitz and Radon [see Eckmann, 1942] gives a strong positive result; that is, a specific number of linearly independent vector fields (varying with the dimension of the sphere) is shown to exist. The solution of the problem was completed by Adams [1962] who showed that these positive results were the best possible.

Before proceeding with further applications, we digress in order to introduce some necessary algebraic ideas. A *directed set* $\varLambda$ is a set with a partial order relation $\leq$ such that given elements a and b in $\varLambda$ there exists an element c in $\varLambda$ with $a \leq c$ and $b \leq c$. A *direct system of sets* is a family of sets $\{X_a\}_{a \in \varLambda}$, where $\varLambda$ is a directed set, and functions

$$f_a^b: \quad X_a \to X_b \qquad \text{whenever} \quad a \leq b,$$

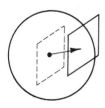

Figure 1.12

satisfying the following requirements:

(i) $f_a^a =$ identity on X_a for each a in Λ;

(ii) if $a \leq b \leq c$, then $f_a^c = f_b^c \circ f_a^b$.

The particular case of interest to us is where the X_a are abelian groups and the f_a^b are homomorphisms. So let $\{X_a, f_a^b\}$ be a direct system of abelian groups and homomorphisms. Define a subgroup R of $\sum_a X_a$ as follows:

$$R = \left\{ \sum_{i=1}^n x_{a_i} \mid \text{there exists a } c \in \Lambda, \ c \geq a_i \text{ for all } i, \text{ and } \sum_{i=1}^n f_{a_i}^c(x_{a_i}) = 0 \right\}.$$

Then the *direct limit* of the system $\{X_a, f_a^b\}$ is the group

$$\varinjlim_a X_a = \sum_a X_a / R.$$

Note that if x_a is in X_a and x_b is in X_b, then they will be equal in the direct limit if for some c in Λ, $c \geq a$ and $c \geq b$ and $f_a^c(x_a) = f_b^c(x_b)$.

1.27 LEMMA Let X be a space and denote by $\{X_a\}$ the family of all compact subsets of X, partially ordered by inclusion. Then the family of groups $\{H_*(X_a)\}$ forms a direct system where the homomorphisms are induced by the inclusion maps. Then

$$\varinjlim_a H_*(X_a) \approx H_*(X).$$

PROOF For each X_a, let the homomorphism

$$g_{a*}: \quad H_*(X_a) \to H_*(X)$$

be induced by the inclusion map. Then set

$$g = \sum_a g_{a*}: \quad \sum_a H_*(X_a) \to H_*(X).$$

Now suppose that $\sum_{i=1}^n x_{a_i}$ is in R; that is, there exists a compact subset $X_b \subseteq X$ such that $X_{a_i} \subseteq X_b$ for each i and

$$\sum_{i=1}^n g_{a_i*}^b(x_{a_i}) = 0 \quad \text{in} \quad H_*(X_b).$$

Then from the commutativity of the diagram

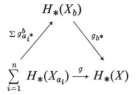

it follows that $g(\sum_{i=1}^{n} x_{a_i}) = 0$ and R is contained in the kernel of g. Thus, g induces a homomorphism

$$\bar{g}: \quad \sum_a H_*(X_a)/R = \varinjlim_a H_*(X_a) \to H_*(X).$$

For any homology class x in $H_n(X)$, represent x by a cycle $\sum n_j \phi_j$. Since σ_n is compact, $\phi_j(\sigma_n)$ is compact in X for each j. Then the chain $\sum n_j \phi_j$ is "supported" on the set $\bigcup_j \phi_j(\sigma_n)$, which is compact since the sum is finite. Thus

$$\bigcup_j \phi_j(\sigma_n) = X_a \quad \text{for some} \quad a,$$

and $\sum n_j \phi_j$ must represent some homology class x_a in $H_n(X_a)$. Moreover, it is evident that $g_{a*}(x_a) = x$; hence, x is in the image of $\bar{g}$ and $\bar{g}$ is an epimorphism.

Now suppose that $\sum_{i=1}^{n} x_{a_i}$ is in $\sum_a H_n(X_a)$ with $g(\sum_{i=1}^{n} x_{a_i}) = 0$. Each x_{a_i} may be represented by a cycle $\sum_j n_{ij} \phi_{ij}$ in X_{a_i}. Then $g(\sum_{i=1}^{n} a_i)$ is represented in X by the cycle $\sum_{i,j} n_{ij} \phi_{ij}$. Since we have assumed that this cycle bounds, there exists an $(n+1)$-chain $\sum_k m_k \psi_k$ in X with $\partial(\sum m_k \psi_k) = \sum_{i,j} n_{ij} \phi_{ij}$. Once again define a subset of X by

$$X_b = \left[\bigcup_k \psi_k(\sigma_{n+1}) \right] \cup \left[\bigcup_i X_{a_i} \right],$$

and note that X_b is compact. Since $\sum m_k \psi_k$ is an $(n+1)$-chain in X_b with $\partial(\sum m_k \psi_k) = \sum_{i,j} n_{ij} \phi_{ij}$, it follows that

$$\sum_{i=1}^{n} g_{a_i\#}^b \left(\sum_j n_{ij} \phi_{ij} \right) \quad \text{is a boundary in} \quad S_n(X_b)$$

and

$$\sum_{i=1}^{n} g_{a_i*}^b(x_{a_i}) = 0 \quad \text{in} \quad H_n(X_b).$$

Thus, $\sum_{i=1}^{n} x_{a_i}$ is in R, $R = $ kernel of g, and $\bar{g}$ is an isomorphism. $\square$

1.28   LEMMA   If $A \subseteq S^n$ is a subset with A homeomorphic to I^k, $0 \leq k \leq n$, then

$$H_j(S^n - A) \approx \begin{cases} Z & \text{for} \quad j = 0 \\ 0 & \text{for} \quad j > 0. \end{cases}$$

PROOF   Proceeding by induction on k, if $k = 0$ then A is a point and $S^n - A$ is homeomorphic to R^n from which the conclusion follows. Assume then that the result is true for $k < m$ and let

$$h: \quad A \to I^m$$

be a homeomorphism. Split the m-cube I^m into its upper and lower halves by setting

$$I^+ = \{(x_1, \ldots, x_m) \in I^m |\ x_1 \geq 0\} \quad \text{and} \quad I^- = \{(x_1, \ldots, x_m) \in I^m |\ x_1 \leq 0\}$$

so that $I^+ \cap I^-$ is homeomorphic to I^{m-1}. For the corresponding decomposition of A denote by $A^+ = h^{-1}(I^+)$ and $A^- = h^{-1}(I^-)$. The set $S^n - (A^+ \cap A^-)$ may be written as the union of two sets $(S^n - A^+) \cup (S^n - A^-)$ satisfying the requirements of the Mayer–Vietoris sequence. So there is an exact sequence

$$H_{j+1}(S^n - (A^+ \cap A^-)) \to H_j(S^n - A) \to H_j(S^n - A^+) \oplus H_j(S^n - A^-)$$
$$\to H_j(S^n - (A^+ \cap A^-)).$$

By the inductive hypothesis, for $j > 0$ the end terms are both zero. This yields an isomorphism

$$H_j(S^n - A) \xrightarrow[i_*^+ \oplus i_*^-]{\approx} H_j(S^n - A^+) \oplus H_j(S^n - A^-).$$

So if $x \in H_j(S^n - A)$ and $x \neq 0$, then either $i_*^+(x) \neq 0$ or $i_*^-(x) \neq 0$. Suppose $i_*^+(x) \neq 0$. Now repeat the procedure by splitting A^+ into two pieces whose intersection is homeomorphic to I^{m-1}. In this manner a sequence of subsets of S^n may be constructed $A = A_1 \supseteq A_2 \supseteq A_3 \supseteq \cdots$ having the property that the inclusion

$$S^n - A \subseteq S_n - A_k$$

induces a homomorphism on homology taking x into a nonzero element of $H_j(S^n - A_k)$, and furthermore that $\bigcap_i A_i$ is homeomorphic to I^{m-1}.

   Now every compact subset of $(S^n - \bigcap_i A_i)$ will be contained in some $(S^n - A_k)$. Thus the isomorphism of Lemma 1.27 factors through the direct limit

$$\lim_{\substack{\longrightarrow \\ k}} H_j(S^n - A_k),$$

so that this direct limit must also be isomorphic to $H_j(S^n - \bigcap_i A_i)$. By the construction, the element of this direct limit represented by x is nonzero; however, by the inductive hypothesis the group $H_j(S^n - \bigcap_i A_i) = 0$. This contradiction implies that no such element x exists and $H_j(S^n - A) = 0$.

For the case $j = 0$, the Mayer–Vietoris sequence yields a monomorphism rather than an isomorphism. If x and y are points in $S^n - A$ with $(x - y) \neq 0$ in $H_0(S^n - A)$, then the above argument may be duplicated to imply that $(x - y)$ must be nonzero in $H_0(S^n - \bigcap_i A_i)$, a contradiction. $\square$

1.29 COROLLARY If $B \subseteq S^n$ is a subset homeomorphic to S^k for $0 \leq k \leq n - 1$, then $H_*(S^n - B)$ is a free abelian group with two generators, one in dimension zero and one in dimension $n - k - 1$.

PROOF Once again inducting on k, note that for $k = 0$, S^k is two points and $S^n - B$ has the homotopy type of S^{n-1}. Since $H_*(S^{n-1})$ satisfies the description, the result is true for $k = 0$. Suppose the result is true for $k - 1$ and write $B = B^+ \cup B^-$, where B^+ and B^- are homeomorphic to closed hemispheres in S^k and $B^+ \cap B^-$ is homeomorphic to S^{k-1}. The Mayer–Vietoris sequence of the covering

$$S^n - (B^+ \cap B^-) = (S^n - B^+) \cup (S^n - B^-)$$

has the form

$$H_{j+1}(S^n - B^+) \oplus H_{j+1}(S^n - B^-) \to H_{j+1}(S^n - (B^+ \cap B^-))$$
$$\to H_j(S^n - B)$$
$$\to H_j(S^n - B^+) \oplus H_j(S^n - B^-).$$

For $j > 0$, both of the end terms are zero by Lemma 1.28. The resulting isomorphism furnishes the inductive step necessary to complete the proof. $\square$

This result may now be applied to prove the following famous theorem.

1.30 THEOREM (*Jordan–Brouwer separation theorem*) An $(n - 1)$-sphere imbedded in S^n separates S^n into two components and it is the boundary of each component.

PROOF   Let $B \subseteq S^n$ be the imbedded copy of S^{n-1}. Then by Corollary 1.29, $H_*(S^n - B)$ is free abelian with two basis elements, both of dimension zero. So $S^n - B$ has two path components. B is closed, so $S^n - B$ is open and hence locally pathwise connected. This implies that the path components are components.

Let C_1 and C_2 be the components of $S^n - B$. Since $C_1 \cup B$ is closed, the boundary of C_1 is contained in B. (Here we mean by the boundary of C_1, the set $\partial C_1 = \bar{C}_1 - C_1°$). The proof will be complete when we show that $B \subseteq \partial C_1$. Let $x \in B$ and U be a neighborhood of x in S^n. Since B is an imbedded copy of S^{n-1}, there is a subset K of $U \cap B$ with $x \in K$ and $B - K$ homeomorphic to D^{n-1} (Figure 1.13).

Now by Lemma 1.28 $H_*(S^n - (B - K)) \approx Z$ with generator in dimension zero. Thus, $S^n - (B - K)$ has one path component. Let $p_1 \in C_1$, $p_2 \in C_2$ and γ a path in $S^n - (B - K)$ between p_1 and p_2. Since C_1 and C_2 are distinct path components in $S^n - B$, the path γ must intersect K. As a result, K contains points of $\bar{C}_1$ and $\bar{C}_2$.

We have shown that an arbitrary neighborhood of x contains points of both $\bar{C}_1$ and $\bar{C}_2$, hence x is in the boundary of C_1 and the proof is complete.   ☐

One final application is the *Brouwer theorem on the invariance of domain*.

1.31   THEOREM   Suppose that U_1 and U_2 are subsets of S^n and that $h: U_1 \to U_2$ is a homeomorphism. Then if U_1 is open, U_2 is also open.

NOTE   It should be observed that this is a nontrivial fact. Of course, it is obviously true if "open" is replaced by "closed," or if the homeomorphism is assumed to be defined over all of S^n. This need not be true in spaces in general. For example, let $W_1 = (\frac{1}{2}, 1]$ and $W_2 = (0, \frac{1}{2}]$ be subsets of $[0, 1]$. If $h: W_1 \to W_2$ is given by $h(x) = x - \frac{1}{2}$, then h is a homeomorphism,

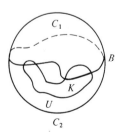

Figure 1.13

W_1 is open, but W_2 is not. It should be evident that there is no extension of h to a homeomorphism of $[0, 1]$ onto itself.

PROOF Suppose $x_2 = h(x_1)$ is some point in U_2. Let V_1 be a neighborhood of x_1 in U_1 with V_1 homeomorphic to D^n and ∂V_1 homeomorphic to S^{n-1}. Set $V_2 = h(V_1)$ and denote by $\partial V_2 = h(\partial V_1)$, so that ∂V_2 is a subset of S^n homeomorphic to S^{n-1} (Figure 1.14).

Then by Lemma 1.28 $S^n - V_2$ is connected, while by Theorem 1.30 $S^n - \partial V_2$ has two components. So $S^n - \partial V_2$ is the disjoint union of $S^n - V_2$ and $V_2 - \partial V_2$, both of which are connected. Hence, they are the components of $S^n - \partial V_2$. This implies that $V_2 - \partial V_2$ is open, contained in U_2, and $x_2 \in V_2 - \partial V_2$. Hence, U_2 is open. $\square$

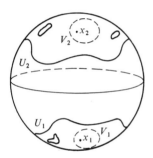

Figure 1.14

chapter 2

ATTACHING SPACES
WITH MAPS

The purpose of this chapter is to develop the basic theory of CW complexes and their homology groups. An equivalence relation on a topological space is seen to produce a new space whose points are the equivalence classes. This gives a means of attaching one space to another via a mapping from a subspace of the first to the second. The case of particular interest is that of attaching a cell to a space via a map defined on the boundary. This leads naturally to the definition of CW complexes. To serve as tools in the study of these spaces, relative homology groups are introduced and the excision theorem is proved. It is shown that the relative groups of adjacent skeletons produce a finitely generated chain complex whose homology is the homology of the space, and this is applied to compute the homology of real projective spaces.

Recall that a relation $\sim$ on a set A is an *equivalence relation* if the following are satisfied:

(i) $a \sim a$,

(ii) $a \sim b \Rightarrow b \sim a$,

(iii) $a \sim b,\ b \sim c \Rightarrow a \sim c$,

for all a, b, and c in A. Such a relation on A gives a decomposition of A into equivalence classes. On the other hand, a decomposition of A into disjoint subsets defines an equivalence relation on A ($a \sim b \Leftrightarrow a$ and b are in the same subset) under which these subsets are the equivalence classes. Denote by $A/\sim$ the set of equivalence classes under $\sim$. By the *quotient function* $\pi: A \rightarrow A/\sim$ we mean the function which assigns to $a \in A$ the equivalence class containing a.

More generally, if $f: A \rightarrow B$ is a function of sets, there is naturally associated an equivalence relation on A. Specifically, $a_1 \sim a_2$ if and only if $f(a_1) = f(a_2)$. In particular if $B = A/\sim$, for some equivalence relation $\sim$, and $f = \pi$, then we recover the original relation $\sim$ in this way.

Now suppose $\sim$ is an equivalence relation on a topological space X. The quotient space $X/\sim$ may be topologized by defining a subset $U \subseteq X/\sim$ to be open if and only if $\pi^{-1}(U)$ is open in X. Note that under this topology, π becomes a continuous function.

Since our main interest is in Hausdorff spaces, we will want to restrict our attention to those equivalence relations on a Hausdorff space X for which the quotient space $X/\sim$ is Hausdorff. For example define an equivalence relation on $[-1, 1]$ by $a \sim -a$ if $|a| < 1$ and $a \sim a$ for all a. Then the images of 1 and -1 in the quotient space cannot be separated by mutually disjoint open sets.

If X is a topological space define

$$D = \{(x, x) \mid x \in X\} \subseteq X \times X$$

the *diagonal* in $X \times X$. Recall that X is Hausdorff if and only if the diagonal is a closed subset of $X \times X$. Now let $\sim$ be an equivalence relation on X and denote by $\varDelta$ the diagonal in $(X/\sim) \times (X/\sim)$. Note that the continuous function

$$\pi \times \pi: \quad X \times X \rightarrow (X/\sim) \times (X/\sim)$$

has

$$(\pi \times \pi)^{-1}(\varDelta) = \{(x, y) \mid x \sim y\}.$$

This subset of $X \times X$ is the *graph of the relation*. The relation $\sim$ on X is *closed* if and only if its graph is a closed subset of $X \times X$. It is evident from the above that if $X/\sim$ is a Hausdorff space, then $\sim$ is a closed relation on X. We now show that the converse is true whenever X is compact.

2.1 PROPOSITION If $\sim$ is a closed relation on a compact Hausdorff space, then $X/\sim$ is Hausdorff.

PROOF Recall that a subset of a compact Hausdorff space is closed if and only if it is compact. Denote by p_1 and p_2 the projection maps of $X \times X$

onto the first and second factors, respectively. Let C be a closed subset of X and $G \subseteq X \times X$ the graph of $\sim$. Then

$$p_2(p_1^{-1}(C) \cap G) = \{y \in X \mid y \sim x \text{ for some } x \in C\}$$
$$= \pi^{-1}(\pi(C)).$$

Now $p_1^{-1}(C) \cap G$ is closed, hence compact, and so $p_2(p_1^{-1}(C) \cap G)$ is compact, hence closed. Thus, for any closed $C \subseteq X$, $\pi^{-1}(\pi(C))$ is closed in X; hence, $\pi(C)$ is closed in $X/\sim$.

If $\bar{x}$ and $\bar{y} \in X/\sim$ are distinct points, then they are closed in $X/\sim$ since they are images of single points in X. Thus, $\pi^{-1}(\bar{x})$ and $\pi^{-1}(\bar{y})$ are disjoint closed subsets of X. Since X is compact Hausdorff, it is normal, and there exist open sets U, V, in X containing $\pi^{-1}(\bar{x})$ and $\pi^{-1}(\bar{y})$, respectively, with $U \cap V = \varnothing$. Let U' and V' be the complements of U and V, so that $\pi(U')$ and $\pi(V')$ are closed subsets of $X/\sim$. Then their complements $X/\sim - \pi(U')$ and $X/\sim - \pi(V')$ are open, disjoint and contain $\bar{x}$ and $\bar{y}$, respectively. Thus, $X/\sim$ is Hausdorff. $\square$

Exercise 1   (a) Give an example of a closed relation $\sim$ on a Hausdorff space X such that $\pi: X \to X/\sim$ is not a closed mapping.

(b) Give an example of a closed relation $\sim$ on a Hausdorff space X such that $X/\sim$ is not Hausdorff.

If a partial relation $\sim'$ is given on a space X, it is possible to associate with $\sim'$ a specific equivalence relation on X. Define an equivalence relation $\sim$ on X by $x \sim y$ if there exists a sequence $x_0, \ldots, x_n$ in X with $x_0 = x$, $x_n = y$, and

(i) $x_{i+1} = x_i$ or
(ii) $x_{i+1} \sim' x_i$ or
(iii) $x_i \sim' x_{i+1}$

for each i. Then $\sim$ is the *equivalence relation generated* by $\sim'$. It is the least equivalence relation that preserves all of the relations from $\sim'$.

For example, let $X = S^n$, $n \geq 1$, and define $\sim$ to be the least equivalence relation on S^n for which $x \sim -x$ for all x. The graph of $\sim$ in $S^n \times S^n$ is the union of the diagonal D and the antidiagonal $D' = \{(x, -x) \mid x \in S^n\}$. This is obviously closed; hence, $S^n/\sim$ is a compact Hausdorff space called *real projective n-space*, $\mathbb{R}P(n)$.

Suppose A, X, and Y are spaces with $A \subseteq X$ and $X \cap Y = \varnothing$. Let $f: A \to Y$ be a continuous function. We consider $X \cup Y$ as a topological space in which X and Y are both open and closed, carrying their original topologies. Let $\sim$ be the least equivalence relation on $X \cup Y$ such that $x \sim f(x)$ for all $x \in A$. The identification space $X \cup Y/\sim$ is the *space obtained by attaching X to Y via f*: $A \to Y$. It is customary to denote $X \cup Y/\sim$ by $X \cup_f Y$.

Exercise 2 Suppose in the above that X and Y are Hausdorff spaces and A is closed in X. Then show that $\sim$ is a closed relation.

2.2 COROLLARY If X and Y are compact Hausdorff spaces, A is closed in X and $f: A \to Y$ is continuous, then $X \cup_f Y$ is a compact Hausdorff space. $\square$

It is not difficult to see that there is a homeomorphic copy of Y sitting in $X \cup_f Y$. We denote by $i: Y \to X \cup_f Y$ the homeomorphism onto this subspace; i may be thought of as the composition of the inclusion of Y in $X \cup Y$ followed by the quotient map $\pi: X \cup Y \to X \cup_f Y$.

A case of particular importance is when $X = D^n$ and $A = S^{n-1} = \partial D^n$. The space $D^n \cup_f Y$ is called the space obtained by *attaching an n-cell to Y via f*. When it may be done without causing confusion, we will denote $D^n \cup_f Y$ by Y_f.

Example Let $X = D^2$, $A = S^1 = \partial D^2$ and Y be a copy of S^1 disjoint from X. Let $f: A \to Y$ be the standard map of degree two given in complex coordinates by $f(e^{i\theta}) = e^{2i\theta}$. The identification space $X \cup_f Y$ is then the *real projective plane*, $\mathbb{R}P(2)$.

The homology groups of this space may be computed by applying the Mayer–Vietoris sequence. In the interior of D^2 pick an open cell U and a

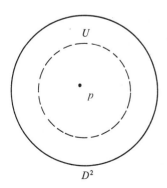

Figure 2.1

point p contained in U (see Figure 2.1). Setting $V = \mathbb{R}\mathrm{P}(2) - \{p\}$, consider the Mayer–Vietoris sequence of the covering $\{U, V\}$. $U \cap V$ and V both have the homotopy type of S^1, whereas U is contractible. In the portion of the sequence given by

$$H_1(U \cap V) \xrightarrow{\alpha} H_1(U) \oplus H_1(V) \xrightarrow{\beta} H_1(\mathbb{R}\mathrm{P}(2))$$
$$\wr\wr \qquad\qquad\qquad \wr\wr$$
$$Z \qquad\qquad\qquad Z$$

it is easy to check that β is an epimorphism. A generating one-cycle in $U \cap V$, when retracted out onto the boundary, is wrapped twice around S^1 since f has degree two. Thus, α is a monomorphism onto $2Z$, and $H_1(\mathbb{R}\mathrm{P}(2)) \approx Z/2Z = Z_2$.

Moreover, the connecting homomorphism

$$H_2(\mathbb{R}\mathrm{P}(2)) \xrightarrow{\Delta} H_1(U \cap V)$$

is a monomorphism whose image is the kernel of α, so $H_2(\mathbb{R}\mathrm{P}(2)) = 0$. All higher-dimensional homology groups are easily seen to be zero, and $\mathbb{R}\mathrm{P}(2)$ is pathwise connected, so its homology is completely determined.

The technique used in this example may easily be adapted to prove the following proposition:

2.3 PROPOSITION   If $f: S^{n-1} \to Y$ is continuous where Y is Hausdorff, then there is an exact sequence

$$\cdots \to H_m(S^{n-1}) \xrightarrow{f_*} H_m(Y) \xrightarrow{i_*} H_m(Y_f) \xrightarrow{\Delta} H_{m-1}(S^{n-1}) \to \cdots$$
$$\to H_0(S^{n-1}) \to H_0(Y) \oplus Z \to H_0(Y_f). \qquad \square$$

This exact sequence shows how closely related are the homology groups of Y and Y_f. If an n-cell has been attached to Y, then $H_n(Y) \xrightarrow{i_*} H_n(Y_f)$ is a monomorphism with cokernel either zero or infinite cyclic. In this sense we may have created a new n-dimensional "hole." On the other hand, $H_{n-1}(Y) \xrightarrow{i_*} H_{n-1}(Y_f)$ is an epimorphism with kernel either zero or cyclic, so the effect of this new n-cell may have been to fill an existing $(n-1)$-dimensional "hole" in Y. Away from these dimensions, the addition of an n-cell does not effect the homology.

Let (X, A) be a pair of spaces and $Y = $ point. Then there is only one map $A \xrightarrow{f} Y$ for $A \neq \varnothing$. The space $X \cup_f Y$ is then denoted by X/A because it can be pictured as the spaced formed from X by collapsing A to a point.

Note that if X is compact Hausdorff and A is closed in X, then X/A is compact Hausdorff.

2.4 PROPOSITION If X and W are compact Hausdorff spaces and $g: X \to W$ is a continuous function onto W such that for some $w_0 \in W$, $g^{-1}(w_0)$ is a closed set $A \subseteq X$, and for $w \neq w_0$, $g^{-1}(w)$ is a single point of X, then W is homeomorphic to X/A.

This follows immediately from the following more general fact.

2.5 PROPOSITION Suppose X, Y, and W are compact Hausdorff spaces and A is a closed subset of X. Let $f: A \to Y$ be continuous and $g: X \cup Y \to W$ continuous and onto. If for each $w \in W$, $g^{-1}(w)$ is either a single point of $X - A$ or the union of a single point $y \in Y$ together with $f^{-1}(y)$ in A, then W is homeomorphic to $X \cup_f Y$.

PROOF If $\pi: X \cup Y \to X \cup_f Y$ is the identification map, g may be factored through π to give a commutative triangle

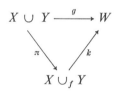

where k is induced by g. Then k is one to one and onto by the properties of g. To see that k is continuous, let C be closed in W. Then $k^{-1}(C)$ is closed if and only if $\pi^{-1}k^{-1}(C)$ is closed. But $\pi^{-1}k^{-1}(C) = g^{-1}(C)$, which is closed since g is continuous. Since $X \cup_f Y$ and W are compact Hausdorff spaces, k is a homeomorphism. $\square$

Example Consider S^{n-1} as the boundary of D^n and let $h_1: D^n - S^{n-1} \to \mathbb{R}^n$ be a homeomorphism. Let $z \in S^n$ and set $h_2: S^n - \{z\} \to \mathbb{R}^n$ to be the homeomorphism given by stereographic projection. Now define a function

$$g: \quad D^n \to S^n \quad \text{by} \quad g(x) = \begin{cases} z & \text{if} \quad x \in S^{n-1} \\ h_2^{-1}h_1(x) & \text{if} \quad x \in D^n - S^{n-1}. \end{cases}$$

Then checking that g satisfies the hypothesis of Proposition 2.4 with $A = S^{n-1}$ we conclude that D^n/S^{n-1} is *homeomorphic to* S^n. Thus, S^n may be viewed as the space given by attaching an n-cell to a point.

Many of the spaces which concern algebraic topologists may be constructed in a similar fashion, that is, by repeatedly attaching cells of varying dimensions to a finite set of points. Before giving a formal definition, we consider a number of important examples.

Example Recall that $RP(n) = S^n/\sim$, wheie $\sim$ is the least equivalence relation on S^n having $x \sim -x$ for all x. Denote by $\pi\colon S^n \to RP(n)$ the quotient map. What space is produced by attaching an $(n+1)$-cell to $RP(n)$ via π?

Regard $S^n \subseteq S^{n+1}$ by identifying $(x_1, \ldots, x_{n+1}) \in S^n$ with $(x_1, \ldots, x_{n+1}, 0) \in S^{n+1}$. This induces an inclusion map $i\colon RP(n) \to RP(n+1)$, of a closed subset. Write S^{n+1} as the union of two subsets $E_+^{n+1} \cup E_-^{n+1}$ which correspond to the upper and lower closed hemispheres, that is, $E_+^{n+1} \cap E_-^{n+1} = S^n$.

There is a homeomorphism $g\colon D^{n+1} \to E_+^{n+1}$. Denote by $f_1\colon D^{n+1} \to RP(n+1)$ the composition of the maps

$$D^{n+1} \xrightarrow{g} E_+^{n+1} \subseteq S^{n+1} \xrightarrow{h} RP(n+1),$$

where h is the quotient map on S^{n+1}.

Thus, we have a mapping of the union

$$D^{n+1} \cup RP(n) \xrightarrow{f_1 \cup i} RP(n+1).$$

It is not difficult to check that $f_1 \cup i$ is onto; in fact, f_1 is onto. Note that for $z \in RP(n+1)$, $f_1^{-1}(z)$ is either a single point of $D^{n+1} - S^n$ or a pair $\{x, -x\}$ in S^n, the latter being true if and only if z lies in the subspace $RP(n)$. Thus, the hypothesis of Proposition 2.5 are satisfied and we conclude that $RP(n+1)$ is *homeomorphic to* $D^{n+1} \cup_\pi RP(n)$, *the space given by attaching an* $(n+1)$-*cell to* $RP(n)$ *via* π.

Suppose X and Y are topological spaces and $x_0 \in X$, $y_0 \in Y$ are base points. In $X \times Y$ there are the subsets $\{x_0\} \times Y$ and $X \times \{y_0\}$. Define $X \vee Y$, the *wedge* of X and Y, to be the union of these two subsets,

$$X \times \{y_0\} \cup \{x_0\} \times Y.$$

Example Denote by I the unit interval in R^1, $\partial I = \{0, 1\}$. The n-cube

$$I^n \subseteq R^n \qquad \text{has} \qquad \partial I^n = \{(x_1, \ldots, x_n) \mid \text{ some } x_i = 0 \text{ or } 1\}.$$

So $I^m \times I^n = I^{m+n}$ and

$$\partial(I^{m+n}) = (\partial I^m \times I^n) \cup (I^m \times \partial I^n).$$

Let $z_m \in S^m$ and $z_n \in S^n$ be base points. There exists a map of pairs $f: (I^m, \partial I^m) \to (S^m, z_m)$, which is a relative homeomorphism; similarly there is a $g: (I^n, \partial I^n) \to (S^n, z_n)$. Taking cartesian products gives a map

$$f \times g: \quad I^m \times I^n \to S^m \times S^n.$$

To see what happens on $\partial(I^m \times I^n)$ note that

$$I^{m+n} - \partial(I^{m+n}) = (I^m - \partial I^m) \times (I^n - \partial I^n).$$

From the properties of f and g, $f \times g$ maps this one to one onto

$$(S^m - z_m) \times (S^n - z_n) = S^m \times S^n - (S^m \times \{z_n\} \cup \{z_m\} \times S^n)$$
$$= S^m \times S^n - S^m \vee S^n.$$

Furthermore, $f \times g$ maps $\partial(I^{m+n})$ onto $S^m \vee S^n$, so by Proposition 2.5, $S^m \times S^n$ is homeomorphic to the space obtained by attaching an $(m + n)$-cell to $S^m \vee S^n$ via the map

$$\partial(I^{m+n}) \approx S^{m+n-1} \to S^m \vee S^n.$$

$S^m \times S^n$ is called a *generalized torus*.

Example   For each integer n identify $\mathbb{R}^{2n}$ with $\mathbb{C}^n$ and denote the points by $(z_1, \ldots, z_n)$. Then $S^{2n-1} \subseteq \mathbb{C}^n$ is given by

$$S^{2n-1} = \{(z_1, \ldots, z_n) \mid \sum |z_i|^2 = 1\}.$$

Define an equivalence relation on S^{2n-1} by $(z_1, \ldots, z_n) \sim (z_1', \ldots, z_n')$ if and only if there exists a complex number λ with $|\lambda| = 1$ such that $z_1' = \lambda z_1, \ldots, z_n' = \lambda z_n$. This is a closed relation, and the space $S^{2n-1}/\sim$ is denoted $\mathbb{C}P(n-1)$, $(n-1)$-*dimensional complex projective space*. [This because its complex dimension is $(n-1)$, real dimension $(2n-2)$.] Recall that in the case of real projective space, a point on the sphere determined a unique real line through the origin and the point was set equivalent to all other points on that line, that is, the antipodal point. In the complex case, a point on the sphere determines a unique complex line through the origin and the point is identified with all other points on that line.

Exercise 3   Let $f: S^{2n-1} \to S^{2n-1}/\sim = \mathbb{C}P(n-1)$ be the identification map. Show that the space formed by attaching a $2n$-cell to $\mathbb{C}P(n-1)$ via f is homeomorphic to $\mathbb{C}P(n)$.

For the case $n = 1$, any two points in S^1 are equivalent; hence, $\mathbb{C}P(0)$ is a point. Now $\mathbb{C}P(1)$ is formed by attaching D^2 to $\mathbb{C}P(0)$, which must yield S^2. Thus, $\mathbb{C}P(1)$ is homeomorphic to S^2. A matter of particular interest is the identification map

$$S^3 \to S^3/\!\sim\; = \mathbb{C}P(1) = S^2.$$

This map

$$h\colon\quad S^3 \to S^2$$

is called the *Hopf map* and is of particular importance in homotopy theory.

Example In the same manner as for the complex number field, we may identify R^4 with the division ring of quaternions by $(x_1, x_2, x_3, x_4) \to x_1 + ix_2 + jx_3 + kx_4$. This identifies R^{4n} with H^n, and the sphere

$$S^{4n-1} = \{(\alpha_1, \ldots, \alpha_n) \in H^n \mid \textstyle\sum |\alpha_i|^2 = 1\}.$$

On S^{4n-1} set $(\alpha_1, \ldots, \alpha_n) \sim (\alpha_1', \ldots, \alpha_n')$ if there exists a $\gamma \in H$ with $|\gamma| = 1$ such that $\alpha_1' = \gamma\alpha_1, \ldots, \alpha_n' = \gamma\alpha_n$. Then $S^{4n-1}/\!\sim$ is $HP(n-1)$, $(n-1)$-*dimensional quaternionic projective space*. As before we find that $HP(0) = pt$, $HP(1) \approx S^4$ and $HP(n)$ is the space given by attaching a $4n$-cell to $HP(n-1)$ via the identification map $S^{4n-1} \to HP(n-1)$. The identification map $h\colon S^7 \to HP(1) = S^4$ is once again called the *Hopf map*.

We now want to compute the homology groups of some of these examples. Leaving the real projective spaces for later in this chapter, first consider the generalized torus $S^m \times S^n$ and assume $m, n \geq 2$.

Recall that $S^m \times S^n$ is given by attaching an $(m+n)$-cell to $S^m \vee S^n$. Denote by $-z_m$ and $-z_n$ the antipodes of the base points $z_m \in S^m$ and $z_n \in S^n$ (see Figure 2.2). Define

$$U = S^m \vee S^n - \{-z_n\} \qquad \text{and} \qquad V = S^m \vee S^n - \{-z_m\}.$$

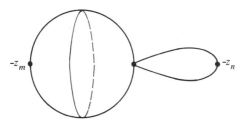

Figure 2.2

Then $\{U, V\}$ gives an open covering of $S^m \vee S^n$, U admits a deformation retraction onto S^m, and V admits a deformation retraction onto S^n. Finally, note that $U \cap V$ has the homotopy type of a point. Thus, in the Mayer–Vietoris sequence for this covering we have

$$H_j(S^m) \oplus H_j(S^n) \approx H_j(S^m \vee S^n) \qquad \text{for} \quad j > 0.$$

Therefore, $H_*(S^m \vee S^n)$ is a free abelian group of rank three having one basis element of dimension zero one of dimension m and one of dimension n.

Now by Proposition 2.3 there is an exact sequence

$$\cdots \to H_i(S^{m+n-1}) \xrightarrow{f_*} H_i(S^m \vee S^n) \to H_i(S^m \times S^n) \to H_{i-1}(S^{m+n-1}) \to \cdots.$$

Since $m, n \geq 2$, $m + n - 1 > m$ and $m + n - 1 > n$. It follows that f_* is the zero map in positive dimensions. On the other hand, if $i = m + n$, the connecting homomorphism

$$H_i(S^m \times S^n) \to H_{i-1}(S^{m+n-1})$$

must be an isomorphism. This information may be combined with special arguments for dimensions zero and one to prove the following:

2.6 PROPOSITION $H_*(S^m \times S^n)$, $m, n \geq 0$, is a free abelian group of rank four having one basis element of each dimension 0, m, n, and $m + n$. $\square$

NOTE This is our first encounter with a nonspherical homology class. Let $\alpha \in H_k(S^k) \approx Z$ be a generator. An homology class $\beta \in H_k(X)$ is *spherical* if there exists a map $f: S^k \to X$ such that $f_*(\alpha) = \beta$. Specifically, if $\beta \in H_2(S^1 \times S^1)$ is a generator, then β is not spherical. Although we are not equipped to prove this at the present time, the basic reason is that β is a product of two one-dimensional homology classes, while $\alpha \in H_2(S^2)$ is not.

Next consider complex projective space $\mathbb{C}P(n)$. For $n = 0, 1$ we know $H_*(\mathbb{C}P(0)) \approx H_*(\text{pt})$ and $H_*(\mathbb{C}P(1)) \approx H_*(S^2)$.

2.7 PROPOSITION

$$H_i(\mathbb{C}P(n)) \approx \begin{cases} Z & \text{for} \quad i = 0, 2, 4, \ldots, 2n \\ 0 & \text{otherwise.} \end{cases}$$

PROOF We proceed by induction on n. From the remarks above, the result is true for $n = 0$ or 1. So suppose it is true for $n - 1 \geq 1$ and recall that $\mathbb{C}P(n)$ may be constructed by attaching a $2n$-cell to $\mathbb{C}P(n-1)$ via the identification map $f: S^{2n-1} \to \mathbb{C}P(n-1)$. By Proposition 2.3 this yields an exact sequence

$$\cdots \to H_i(S^{2n-1}) \xrightarrow{f_*} H_i(\mathbb{C}P(n-1)) \xrightarrow{j_*} H_i(\mathbb{C}P(n)) \xrightarrow{\Delta} H_{i-1}(S^{2n-1}) \to \cdots$$

for $i > 0$. For strictly algebraic reasons the homomorphism f_* must be zero in positive dimensions. So for $i > 1$, this gives a collection of short exact sequences

$$0 \to H_i(\mathbb{C}P(n-1)) \xrightarrow[j_*]{} H_i(\mathbb{C}P(n)) \xrightarrow[\Delta]{} H_{i-1}(S^{2n-1}) \to 0.$$

These, together with the induction hypothesis, the fact that j_* is an epimorphism in dimension one and the fact that $\mathbb{C}P(n)$ is pathwise connected, complete the inductive step and the result follows. $\square$

2.8 PROPOSITION

$$H_i(H\mathrm{P}(n)) \approx \begin{cases} Z & \text{for } i = 0, 4, 8, \ldots, 4n \\ 0 & \text{otherwise.} \end{cases}$$

PROOF   The proof of this is entirely analogous to that for Proposition 2.7. $\square$

With these examples in mind we now develop some of the basic properties of spaces constructed in this way. To do so it is necessary to introduce the *relative homology groups*, a useful generalization initiated by Lefschetz in the 1920s. The concept is entirely analogous to that of the quotient of a group by a subgroup. If A is a subspace of X then we set two chains of X equal modulo A if their difference is a chain in A. In particular a chain in X is a cycle modulo A if its boundary is contained in A. This reflects the structure of $X - A$ and the way that it is attached to A. In a sense, changes in the interior of A, away from its boundary with $X - A$, should not alter these homology groups.

To introduce the necessary homological algebra, let $C = \{C_n, \partial\}$ be a chain complex. $D = \{D_n, \partial\}$ is a *subcomplex* of C if $D_n \subseteq C_n$ for each n and the boundary operator for D is the restriction of the boundary operator for C. Define the *quotient chain complex*

$$C/D = \{C_n/D_n, \partial'\},$$

where $\partial'\{c\} = \{\partial c\}$ for $\{c\}$ the coset containing c. For convenience the prime will be omitted and all boundary operators will continue to be denoted by ∂.

There is a natural short exact sequence of chain complexes and chain maps

$$0 \to D \xrightarrow{i} C \xrightarrow{\pi} C/D \to 0,$$

where i is the inclusion and π is the projection. From Theorem 1.13 this leads to a long exact sequence of homology groups

$$\cdots \to H_n(D) \xrightarrow{i_*} H_n(C) \xrightarrow{\pi_*} H_n(C/D) \xrightarrow{\Delta} H_{n-1}(D) \xrightarrow{i_*} \cdots.$$

For clarity denote by $\{\ \}$ the equivalence relation in C/D and by $\langle\ \rangle$ the equivalence relation in homology.

To see how the connecting homomorphism Δ is defined let $\{c\}$ be a cycle in $Z_n(C/D)$. To determine $\Delta(\langle\{c\}\rangle)$, represent $\{c\}$ by an element $c \in C_n$ having $\partial c \in D_{n-1}$. Of course, $\partial c \in Z_{n-1}(D)$ and hence represents a class in $H_{n-1}(D)$. Thus, we have

$$\Delta(\langle\{c\}\rangle) = \langle\partial c\rangle.$$

More generally, if $E \subseteq D \subseteq C$ are chain complexes and subcomplexes, there is a short exact sequence of chain complexes and chain maps

$$0 \to D/E \to C/E \to C/D \to 0.$$

In the corresponding long exact homology sequence

$$\cdots \to H_n(D/E) \to H_n(C/E) \to H_n(C/D) \xrightarrow{\Delta'} H_{n-1}(D/E) \to \cdots$$

the connecting homomorphism is given by $\Delta'(\langle\{c\}\rangle) = \langle\{\partial c\}\rangle$, which may be viewed as the composition

$$H_n(C/D) \xrightarrow{\Delta} H_{n-1}(D) \xrightarrow{\pi_*} H_{n-1}(D/E).$$

This is natural in the sense that if $E' \subseteq D' \subseteq C'$ are chain complexes and subcomplexes and $f: C \to C'$ is a chain map for which $f(D) \subseteq D'$ and $f(E) \subseteq E'$, then the induced homomorphisms on homology groups give a transformation between the long exact homology sequences in which each rectangle commutes.

By a *pair of spaces* (X, A) we mean a space X together with a subspace $A \subseteq X$. If (X, A) is a pair of spaces, $S_*(A)$ may be viewed as a subcomplex of $S_*(X)$. The *singular chain complex* of X mod A is defined by

$$S_*(X, A) = S_*(X)/S_*(A).$$

The homology of this chain complex, the *relative singular homology of X mod A*, is thus given by

$$H_n(X, A) = H_n(S_*(X)/S_*(A)).$$

From the previous observations any pair (X, A) has an exact homology sequence

$$\cdots \to H_n(A) \xrightarrow{i_*} H_n(X) \xrightarrow{\pi_*} H_n(X, A) \xrightarrow{\Delta} H_{n-1}(A) \to \cdots.$$

In this sense $H_*(X, A)$ is a measure of how far $i_*: H_*(A) \to H_*(X)$ is from being an isomorphism. That is, i_* is an isomorphism of graded groups if and only if $H_*(X, A) = 0$. Thus, we have immediately the following:

2.9 PROPOSITION If (X, A) is a pair for which A is a deformation retract of X, then $H_*(X, A) = 0$. $\square$

More generally if (X, A, B) is a triple of spaces, that is, $B \subseteq A \subseteq X$, there results a short exact sequence of chain complexes

$$0 \to S_*(A, B) \to S_*(X, B) \to S_*(X, A) \to 0$$

which yields the corresponding long exact sequence of relative homology groups. It is conventional to define $S_*(\varnothing) = 0$ so that $H_*(X, \varnothing) = H_*(X)$ and all homology groups may be viewed as groups of pairs.

Given pairs (X, A) and (Y, B) a *map of pairs* $f: (X, A) \to (Y, B)$ is a continuous function $f: X \to Y$ for which $f(A) \subseteq B$. Note that for such a map $f_\#(S_*(A)) \subseteq S_*(B)$, so that there is associated a homomorphism

$$f_\#: \quad S_*(X, A) \to S_*(Y, B)$$

which is a chain map, hence also a homomorphism on the relative homology groups. Note that the homomorphisms of degree zero in the exact sequence of a triple are induced by the inclusion maps of pairs.

Two maps of pairs $f, g: (X, A) \to (Y, B)$ are *homotopic as maps of pairs* if there exists a map of pairs

$$F: \quad (X \times I, A \times I) \to (Y, B)$$

such that $F(x, 0) = f(x)$ and $F(x, 1) = g(x)$. Note that this says that in continuously deforming f into g, it is required that at each stage we map A into B.

2.10 THEOREM If $f, g: (X, A) \to (Y, B)$ are homotopic as maps of pairs, then $f_* = g_*$ as homomorphisms from $H_*(X, A)$ to $H_*(Y, B)$.

PROOF As before define $i_0, i_1: (X, A) \to (X \times I, A \times I)$ by $i_0(x) = (x, 0)$ and $i_1(x) = (x, 1)$ and note that it is sufficient to show that $i_{0\#}$ and $i_{1\#}$ are chain homotopic.

Using the same technique as for the absolute case of Theorem 1.10 we construct a natural homomorphism

$$T: \quad S_n(X) \to S_{n+1}(X \times I)$$

having

$$\partial T + T\partial = i_{0\#} - i_{1\#}$$

and observe that the restriction of T has $T(S_n(A)) \subseteq S_{n+1}(A \times I)$. Thus, there is induced the desired chain homotopy

$$T: \quad S_n(X, A) \to S_{n+1}(X \times I, A \times I). \quad \square$$

Example To illustrate the difference between maps being absolutely homotopic and homotopic as maps of pairs, consider the following example. Let $X = [0, 1]$, $A = \{0, 1\}$, and $Y = S^1$, $B = \{1\}$. Define

$$g, f: \quad X \to Y$$

by $f(x) = e^{2\pi i x}$ and $g(x) = 1$. Then f and g are maps of pairs $(X, A) \to (Y, B)$ and f and g are absolutely homotopic as maps from X to Y but they are not homotopic as maps of pairs.

Exercise 4 (*The five lemma*) Suppose that

$$
\begin{array}{ccccccccc}
C_1 & \xrightarrow{\alpha_1} & C_2 & \xrightarrow{\alpha_2} & C_3 & \xrightarrow{\alpha_3} & C_4 & \xrightarrow{\alpha_4} & C_5 \\
\downarrow{\scriptstyle f_1} & & \downarrow{\scriptstyle f_2} & & \downarrow{\scriptstyle f_3} & & \downarrow{\scriptstyle f_4} & & \downarrow{\scriptstyle f_5} \\
D_1 & \xrightarrow{\beta_1} & D_2 & \xrightarrow{\beta_2} & D_3 & \xrightarrow{\beta_3} & D_4 & \xrightarrow{\beta_4} & D_5
\end{array}
$$

is a diagram of abelian groups and homomorphisms in which the rows are exact and each square is commutative. Then show

(i) if f_2, f_4 are epimorphisms and f_5 is a monomorphism, then f_3 is an epimorphism;

(ii) if f_2, f_4 are monomorphisms and f_1 is an epimorphism, then f_3 is a monomorphism.

Note that, as a special case of this exercise, if f_1, f_2, f_4, and f_5 are isomorphisms, then f_3 is an isomorphism.

As pointed out before it seems that those points of A which are not close to the complement of A in X (see Figure 2.3) make no contribution to the relative homology group of the pair (X, A). This property is formally set forth in the following *excision* theorem.

2.11   THEOREM   If (X, A) is a pair of spaces and U is a subset of A with $\bar{U}$ contained in the interior of A, then the inclusion map

$$i:\ (X - U, A - U) \to (X, A)$$

induces an isomorphism on relative homology groups

$$i_*:\ H_*(X - U, A - U) \to H_*(X, A).$$

That is, such a set U may be excised without altering the relative homology groups.

PROOF   Denote by $\mathfrak{U}$ the covering of X given by the two sets $X - U$ and Int A. By assumption their interiors cover X; thus, their interiors also cover A and we set $\mathfrak{U}'$ to be the covering of A given by $\{A - U,\ \text{Int } A\}$. Then by Theorem 1.14 the inclusion homomorphisms of chains

$$i:\ S_*^{\mathfrak{U}}(X) \to S_*(X) \qquad \text{and} \qquad i':\ S_*^{\mathfrak{U}'}(A) \to S_*(A)$$

both induce isomorphisms on homology.

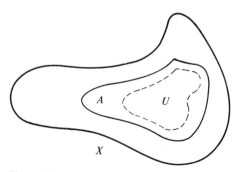

Figure 2.3

Considering $S_*^{\mathfrak{U}'}(A)$ as a subcomplex of $S_*^{\mathfrak{U}}(X)$ there is a chain mapping of chain complexes

$$j: \quad S_*^{\mathfrak{U}}(X)/S_*^{\mathfrak{U}'}(A) \to S_*(X)/S_*(A) = S_*(X, A).$$

The chain mappings i, i', and j give rise to the following diagram of homology groups

$$\cdots \to H_n(S_*^{\mathfrak{U}'}(A)) \to H_n(S_*^{\mathfrak{U}}(X)) \to H_n(S_*^{\mathfrak{U}}(X)/S_*^{\mathfrak{U}'}(A)) \to H_{n-1}(S_*^{\mathfrak{U}'}(A)) \to \cdots$$

$$\cdots \longrightarrow H_n(A) \longrightarrow H_n(X) \longrightarrow H_n(X, A) \longrightarrow H_{n-1}(A) \longrightarrow \cdots$$

with vertical maps i_*', i_*, j_*, i_*'.

Since i_*' and i_* are isomorphisms, it follows from the five lemma (see Exercise 4) that j_* is an isomorphism.

Now we can write $S_*^{\mathfrak{U}}(X)$ as the sum of two subgroups

$$S_*^{\mathfrak{U}}(X) = S_*(X - U) + S_*(\text{Int } A),$$

but not necessarily as a direct sum. Similarly

$$S_*^{\mathfrak{U}'}(A) = S_*(A - U) + S_*(\text{Int } A).$$

Then by elementary group theory

$$S_*^{\mathfrak{U}}(X)/S_*^{\mathfrak{U}'}(A) \approx S_*(X - U)/S_*(A - U) = S_*(X - U, A - U).$$

Composing this isomorphism with the chain map j there is induced on homology the desired isomorphism

$$H_*(X - U, A - U) \to H_*(X, A). \qquad \square$$

A short exact sequence of abelian groups and homomorphisms

$$0 \to A \xrightarrow{f} B \xrightarrow{g} C \to 0$$

is *split exact* if $f(A)$ is a direct summand of B.

Exercise 5   Suppose $0 \to A \xrightarrow{f} B \xrightarrow{g} C \to 0$ is short exact. Then the following are equivalent:

   (i)   the sequence is split exact;
   (ii) there exists a homomorphism $\bar{f}: B \to A$ with $\bar{f} \circ f =$ identity;
   (iii) there exists a homomorphism $\bar{g}: C \to B$ with $g \circ \bar{g} =$ identity.

Let X be a space and y a single point. Denote by $\alpha\colon X \to y$ the map of X into y. Then there is the induced homomorphism on homology

$$\alpha_*\colon \quad H_*(X) \to H_*(y).$$

Denote the kernel of α_* by $\tilde{H}_*(X)$. This subgroup of $H_*(X)$ is the *reduced homology group* of X. Note that since $H_i(y) = 0$ for $i \neq 0$, $\tilde{H}_i(X) = H_i(X)$ for $i \neq 0$. Furthermore, if $X \neq \varnothing$, then α_* is an epimorphism so that $\tilde{H}_0(X)$ is free abelian with one fewer basis element than $H_0(X)$. Note that if $f\colon X \to Y$ is a map, then $f_*\colon \tilde{H}_*(X) \to \tilde{H}_*(Y)$. For example, $\tilde{H}_*(S^n)$ is free abelian with one basis element in dimension n.

2.12 PROPOSITION   If $x_0 \in X$, then $H_*(X, x_0) \approx \tilde{H}_*(X)$.

PROOF   In the exact homology sequence of the pair (X, x_0) the homomorphism $H_i(x_0) \to H_i(X)$ is a monomorphism for each i. Thus, the long sequence breaks up into a collection of short exact sequences:

$$0 \to H_i(x_0) \xrightarrow{\ i_*\ } H_i(X) \xrightarrow{\ j_*\ } H_i(X, x_0) \to 0.$$

The map $\alpha\colon X \to x_0$ induces $\alpha_*\colon H_i(X) \to H_i(x_0)$ which splits the sequence. Thus there is a homomorphism $\beta\colon H_i(X, x_0) \to H_i(X)$ with $j_*\beta = $ identity. This β is then an isomorphism onto the subgroup $\tilde{H}_i(X)$.   □

A subspace A of a space X is a *strong deformation retract* of X if there exists a map $F\colon X \times I \to X$ such that

(i) $F(x, 0) = x$ for all $x \in X$;
(ii) $F(x, 1) \in A$ for all $x \in X$;
(iii) $F(a, t) = a$ for all $a \in A$ and $t \in I$.

Exercise 6   Let X be the space given by the unit interval together with a family of segments approaching it as pictured in Figure 2.4. If A is the unit

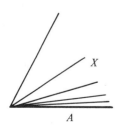

Figure 2.4

interval, show that A is a deformation retract of X but not a strong deformation retract.

2.13 PROPOSITION Let (X, A) be a pair in which X is compact Hausdorff, A is closed in X and A is a strong deformation retract of X. Let $\pi: X \to X/A$ be the identification map and denote by y the point $\pi(A)$ in X/A. Then $\{y\}$ is a strong deformation retract of X/A.

PROOF Denote by $F: X \times I \to X$ the map given by the fact that A is a strong deformation retract of X. We must exhibit a map $\tilde{F}: (X/A) \times I \to X/A$ having $\tilde{F}(\tilde{x}, 0) = \tilde{x}$, $\tilde{F}(\tilde{x}, 1) = y$ for all $\tilde{x} \in X/A$ and $\tilde{F}(y, t) = y$ for all $t \in I$. Thus, it would be sufficient to define a map so that the following diagram is commutative:

$$
\begin{array}{ccc}
X \times I & \xrightarrow{\ \ F\ \ } & X \\
{\scriptstyle \pi \times \mathrm{id}} \downarrow & & \downarrow {\scriptstyle \pi} \\
(X/A) \times I & \xrightarrow{\ \ \tilde{F}\ \ } & X/A
\end{array}
$$

So define $\tilde{F} = \pi \circ F \circ (\pi \times \mathrm{id})^{-1}$. To see that this is single valued, let $(\tilde{x}, t) \in (X/A) \times I$. Then

$$(\pi \times \mathrm{id})^{-1}(\tilde{x}, t)$$

is just (x, t) if $x \notin A$ and is $A \times \{t\}$ if $x \in A$. So if $x \notin A$, this is obviously single valued. If $x \in A$, note that $F(A \times \{t\}) \subseteq A$ and $\pi(A) = y$. Hence, $\tilde{F}$ is single valued.

To show that $\tilde{F}$ is continuous, let $C \subseteq X/A$ be a closed set. Then $F^{-1} \circ \pi^{-1}(C)$ is closed in $X \times I$, hence compact. Thus, $(\pi \times \mathrm{id}) \circ F^{-1} \circ \pi^{-1}(C)$ is compact in $X/A \times I$, hence closed. Therefore, $\tilde{F}$ is continuous. $\quad\square$

2.14 THEOREM Let (X, A) be a pair with X compact Hausdorff and A closed in X, where A is a strong deformation retract of some closed neighborhood of A in X. Let $\pi: (X, A) \to (X/A, y)$ be the identification map. Then

$$\pi_*: \quad H_*(X, A) \to H_*(X/A, y)$$

is an isomorphism.

PROOF Let U be a compact neighborhood of A in X which admits a strong deformation retraction onto A (see Figure 2.5). Applying Proposition 2.13

to the pair (U, A) we observe that $\{y\}$ is a strong deformation retract of $\pi(U)$. Thus, in the exact sequence of the triple $(X/A, \pi(U), y)$,

$$\cdots \to H_n(\pi(U), y) \to H_n(X/A, y) \to H_n(X/A, \pi(U)) \to H_{n-1}(\pi(U), y) \to \cdots$$

it follows that $H_*(\pi(U), y) = 0$. Hence, the inclusion map of pairs induces an isomorphism

$$H_*(X/A, y) \approx H_*(X/A, \pi(U)).$$

Recall that since X is compact Hausdorff, it is also normal. Now Int U is an open set containing the closed set A, so there exists an open set V with $A \subseteq V$ and $\bar{V} \subseteq$ Int U. Thus, V may be excised from the pair (X, U) to induce an isomorphism

$$H_*(X - V, U - V) \approx H_*(X, U).$$

Since A is a strong deformation retract of U, it follows from the exact sequence that

$$H_*(X, A) \approx H_*(X, U).$$

These two isomorphisms may be combined to give

$$H_*(X, A) \approx H_*(X - V, U - V).$$

In similar fashion the set $\pi(V)$ may be excised from the pair $(X/A, \pi(U))$ to give an isomorphism

$$H_*(X/A, y) \approx H_*(X/A, \pi(U)) \approx H_*(X/A - \pi(V), \pi(U) - \pi(V)).$$

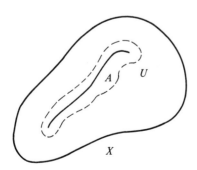

Figure 2.5

Now note that since V is a neighborhood of the set A which was collapsed, the restriction of the map π gives a homeomorphism of pairs

$$\pi: \quad (X - V, U - V) \rightarrow (X/A - \pi(V), \pi(U) - \pi(V)),$$

and so an isomorphism of their homology groups. All of these combine to give the desired isomorphism

$$H_*(X, A) \approx H_*(X/A, y). \quad \square$$

2.15 COROLLARY If (X, A) is a compact Hausdorff pair for which A is a strong deformation retract of some compact neighborhood of A in X, then

$$H_*(X, A) \approx \tilde{H}_*(X/A). \quad \square$$

If $f: (X, A) \rightarrow (Y, B)$ is a map of pairs such that f maps $X - A$ one to one and onto $Y - B$, then f is a *relative homeomorphism*. Under certain conditions on the pairs a relative homeomorphism will induce an isomorphism of relative homology groups.

2.16 THEOREM (*Relative homeomorphism theorem*) If $f: (X, A) \rightarrow (Y, B)$ is a relative homeomorphism of compact Hausdorff pairs in which A is a strong deformation retract of some compact neighborhood in X and B is a strong deformation retract of some compact neighborhood in Y, then

$$f_*: \quad H_*(X, A) \rightarrow H_*(Y, B) \quad \text{is an isomorphism.}$$

PROOF Consider the diagram of spaces and maps, where π and π' are the identification maps and $f' = \pi' \circ f \circ \pi^{-1}$:

$$
\begin{array}{ccc}
X & \xrightarrow{f} & Y \\
\downarrow{\scriptstyle \pi} & & \downarrow{\scriptstyle \pi'} \\
X/A & \xrightarrow{f'} & Y/B
\end{array}
$$

As in the proof of Proposition 2.13 it is easy to see that f' is single valued and continuous. Since f is a relative homeomorphism, f' is one to one and onto. But X/A and Y/B are compact Hausdorff spaces, so f' is a homeomorphism.

Denoting $x_0 = \pi(A)$ and $y_0 = \pi'(B)$ there is the corresponding diagram of relative homology groups and induced homomorphisms:

$$
\begin{array}{ccc}
H_*(X, A) & \xrightarrow{\ f_*\ } & H_*(Y, B) \\
\downarrow{\scriptstyle \pi_*} & & \downarrow{\scriptstyle \pi_*'} \\
H_*(X/A, x_0) & \xrightarrow{\ f_*'\ } & H_*(Y/B, y_0)
\end{array}
$$

By Theorem 2.14 the homomorphisms π_* and π_*' are isomorphisms. Also f_*' is an isomorphism since f' is a homeomorphism. Thus

$$f_*:\ H_*(X, A) \to H_*(Y, B) \quad \text{is an isomorphism.} \qquad \square$$

Examples (1) There is a relative homeomorphism

$$f:\ (D^n, S^{n-1}) \to (S^n, z),$$

where z is any point in S^n. Both pairs satisfy the hypotheses of Theorem 2.16, so there is an isomorphism

$$f_*:\ H_*(D^n, S^{n-1}) \to H_*(S^n, z) \approx \tilde{H}_*(S^n).$$

(2) To see that the hypotheses of the theorem are actually necessary, consider the following example. Using the curve $\sin(1/x)$ construct a space as shown in Figure 2.6a, where X is the curve together with those points "inside," and A is the boundary. Let $Y = D^2$ and $B = \partial D^2 = S^1$ (Figure 2.6b). Then (X, A) and (Y, B) are compact Hausdorff pairs. By flattening the pathological part of A it is possible to define a map of pairs $f: (X, A) \to (Y, B)$ which is a relative homeomorphism. However, it cannot induce an isomorphism on homology because $H_2(X, A) = 0$ and $H_2(Y, B) \approx Z$.

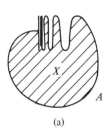

(a)

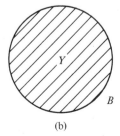

(b)

Figure 2.6

The result fails because A is not a strong deformation retract of some compact neighborhood of A in X.

The fact that $H_2(X, A) = 0$ is an easy consequence of the exact sequence of the pair (X, A),

$$\cdots \to H_2(A) \to H_2(X) \to H_2(X, A) \to H_1(A) \to \cdots.$$

Now X is contractible, so $H_2(X) = 0$. On the other hand, if $\sum n_i \phi_i$ is a 1-chain in A, the sum must be finite. Since the curve A is not locally connected, the union of the images of these singular simplices cannot bridge the gap in the $\sin(1/x)$ curve. Thus, the chain is supported by some contractible subset of A, so that if it is a cycle, it is also a boundary. Therefore, $H_1(A) = 0$ and by exactness $H_2(X, A) = 0$.

2.17   LEMMA   Let $f: S^{n-1} \to Y$ be a map, where Y is a compact Hausdorff space. If Y_f is the space obtained by attaching an n-cell to Y via f, then Y is a strong deformation retract of some compact neighborhood of Y in Y_f.

PROOF   Let $U \subseteq D^n$ be the subset given by $U = \{x \in D^n \mid \| x \| \geq \frac{1}{2}\}$, and observe that U is a compact neighborhood of S^{n-1} in D^n. Define a map $F: (U \cup Y) \times I \to U \cup Y$ by

$$F(x, t) = \begin{cases} x & \text{if } x \in Y \\ (1 - t)x + t \cdot \dfrac{x}{\| x \|} & \text{if } x \in U. \end{cases}$$

Then F is continuous, $F(x, 0) = x$ and $F(x, 1) \in S^{n-1} \cup Y$ for all x, and if $x \in S^{n-1} \cup Y$, then $F(x, t) = x$ for all t. Thus, F is a strong deformation retraction of $U \cup Y$ onto $S^{n-1} \cup Y$.

Now let $\pi: D^n \cup Y \to Y_f$ be the identification map and consider the diagram

$$\begin{array}{ccc} (U \cup Y) \times I & \xrightarrow{\;F\;} & U \cup Y \\ \big\downarrow{\scriptstyle \pi \times \mathrm{id}} & & \big\downarrow{\scriptstyle \pi} \\ \pi(U \cup Y) \times I & \xrightarrow[\;F'\;]{} & \pi(U \cup Y) \end{array}$$

As before define

$$F' = \pi \circ F \circ (\pi \times \mathrm{id})^{-1}.$$

Then F' is well defined, continuous and gives a strong deformation retraction of the compact neighborhood $\pi(U \cup Y)$ of $\pi(Y)$ onto $\pi(Y)$. $\square$

NOTE   Denote by h the composition

$$D^n \xrightarrow{\text{incl}} D^n \cup Y \xrightarrow{\pi} Y_f.$$

Then h gives a map of pairs $h: (D^n, S^{n-1}) \to (Y_f, Y)$ which is a relative homeomorphism. The hypotheses of Lemma 2.17 and Theorem 2.16 are satisfied, so we may conclude that

$$h_*: \quad H_*(D^n, S^{n-1}) \to H_*(Y_f, Y)$$

is an isomorphism. Therefore, $H_*(Y_f, Y)$ is a free abelian group on one basis element of dimension n.

Suppose that $D_1^n, \ldots, D_k^n$ is a finite number of disjoint n-cells with boundaries $S_1^{n-1}, \ldots, S_k^{n-1}$. For each $i = 1, \ldots, k$ let $f_i: S_i^{n-1} \to Y$ be a map into a fixed space Y. Define $\sim$ to be the least equivalence relation on $D_1^n \cup \cdots \cup D_k^n \cup Y$ for which $x_i \sim f_i(x_i)$ whenever $x_i \in S_i^{n-1}$.

Then $D_1^n \cup \cdots \cup D_k^n \cup Y/\sim$ may be denoted $Y_{f_1, \ldots, f_k}$, the space obtained by attaching n-cells to Y via $f_1, \ldots, f_k$.

Conversely, if (X, Y) is a compact Hausdorff pair for which there exists a relative homeomorphism

$$F: \quad (D_1^n \cup \cdots \cup D_k^n, S_1^{n-1} \cup \cdots \cup S_k^{n-1}) \to (X, Y),$$

then X is homeomorphic to $Y_{f_1, \ldots, f_k}$ where $f_i = F|_{S_i^{n-1}}$.

A *finite CW complex* is a compact Hausdorff space X and a sequence $X^0 \subseteq X^1 \subseteq \cdots \subseteq X^n = X$ of closed subspaces such that

   (i) X^0 is a finite set of points;
   (ii) X^k is homeomorphic to a space obtained by attaching a finite number of k-cells to X^{k-1}.

Note that $X^k - X^{k-1}$ is thus homeomorphic to a finite disjoint union of open k-cells, denoted $E_1^k, \ldots, E_{r_k}^k$. These are the k-*cells* of X. Using the convention that $D^0 = \text{point}$ and $\partial D^0 = S^{-1} = \varnothing$, the requirements (i) and (ii) may be replaced by the condition that for each k there exist a relative homeomorphism

$$f: \quad (D_1^k \cup \cdots \cup D_{r_k}^k, S_1^{k-1} \cup \cdots \cup S_{r_k}^{k-1}) \to (X^k, X^{k-1}).$$

It is easy to verify that the cells of X have the following properties:

   (a) $\{E_i^k \mid k = 0, 1, \ldots, n; i = 1, \ldots, r_k\}$ is a partition of X into disjoint sets;

(b) for each k and i the set $\bar{E}_i^k - E_i^k$ is contained in the union of all cells of lower dimension;

(c) $X^k = \bigcup_{k' \le k} E_j^{k'}$;

(d) for each i and k there exists a relative homeomorphism

$$h: \quad (D^k, S^{k-1}) \to (\bar{E}_i^k, \bar{E}_i^k - E_i^k).$$

These properties characterize finite CW complexes and will be used as an alternate definition whenever it is convenient. The closed subset X^k is the *k-skeleton* of X. If $X^n = X$ and $X^{n-1} \ne X$, then X is *n-dimensional*.

Example It should be evident that for a given space there may be many different decompositions into cells and skeletons (see Figure 2.7). For example, let $X = S^2$. If z is a point in S^2, then S^2 may be described as the space obtained by attaching a 2-cell to z. This gives S^2 a cell structure in which there is one 0-cell and one 2-cell (Figure 2.7a).

If z' is another point in S^2 and α is a simple path from z to z', we have a cell structure with two 0-cells, a 1-cell, and a 2-cell (Figure 2.7b). Why was it necessary to include the 1-cell α when two vertices were used?

Further cells may be included as shown in the third figure, in which there are two 0-cells, three 1-cells, and three 2-cells (Figure 2.7c). While there is considerable freedom in assigning a cell structure to a finite CW complex, it is apparent that any change in the number of cells in a certain dimension dictates some corresponding change in the number of cells in other dimensions.

Note that one apparent advantage CW complexes have over simplicial complexes is that considerably fewer cells are generally necessary in the decomposition of a complex.

2.18 PROPOSITION If X and Y are finite CW complexes, then $X \times Y$ is a finite CW complex in a natural way.

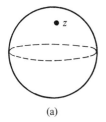   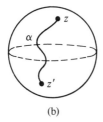   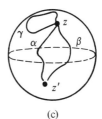

      (a)   (b)   (c)

Figure 2.7

PROOF   Suppose the cellular decompositions of X and Y are given by $\{E_i^k\}$ and $\{E_j'^l\}$. The obvious candidate for a cellular decomposition for $X \times Y$ is the collection $\{E_i^k \times E_j'^l\}$. First note that this is a partition of $X \times Y$ into a finite number of sets homeomorphic to open cells. Also

$$\overline{E_i^k \times E_j'^l} - E_i^k \times E_j'^l = \overline{E_i^k} \times \overline{E_j'^l} - E_i^k \times E_j'^l$$
$$= (\overline{E_i^k} - E_i^k) \times \overline{E_j'^l} \cup \overline{E_i^k} \times (\overline{E_j'^l} - E_j'^l),$$

which is contained in the union of all cells of dimension less than $k + l$.

To check the third requirement we may assume that there are relative homeomorphisms

$$f\colon \ (I^k, \partial I^k) \to (\bar{E}_i^k, \bar{E}_i^k - E_i^k) \qquad \text{and} \qquad g\colon \ (I^l, \partial I^l) \to (\bar{E}_j'^l, \bar{E}_j'^l - E_j'^l).$$

Then $f \times g\colon (I^{k+l}, \partial I^{k+l}) \to \overline{(E_i^k \times E_j'^l, \ \overline{E^k \times E_j'^l}} - E_i^k \times E_j'^l)$ gives the desired relative homeomorphism. $\square$

Examples   (1) Taking the decomposition of S^1 into one 0-cell (z) and one 1-cell (α) as in Figure 2.8a, the torus $S^1 \times S^1$ is naturally given the decomposition into one 0-cell $(z \times z)$, two 1-cells $(z \times \alpha$ and $\alpha \times z)$ and one 2-cell $(\alpha \times \alpha)$ (Figure 2.8b).

(2) Recall that $\mathbb{R}P(0) = \text{pt}$ and $\mathbb{R}P(k)$ is obtained by attaching a k-cell to $\mathbb{R}P(k - 1)$. Thus, $\mathbb{R}P(n)$ is an n-dimensional finite CW complex with one cell in each dimension $0, \ldots, n$. Moreover, the k-skeleton of $\mathbb{R}P(n)$ under this structure is just $\mathbb{R}P(k)$.

(3) Similarly $\mathbb{C}P(n)$ is a finite CW complex of dimension $2n$ with one cell in each even dimension, $0, 2, 4, \ldots, 2n$. Also $\mathbb{C}P(k) = $ the $2k$-skeleton of $\mathbb{C}P(n) = $ the $(2k + 1)$-skeleton of $\mathbb{C}P(n)$ for $0 \leq k \leq n$. An anologous structure may be given to quaternionic projective space.

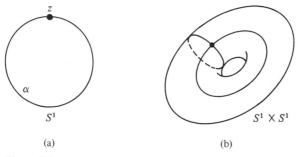

(a)   (b)

Figure 2.8

If X is a finite CW complex with cells $\{E_i^k\}$, then a subset A of X is a *subcomplex* of X if whenever $A \cap E_i^k \neq \varnothing$ then $\bar{E}_i^k \subseteq A$. Note that if A is a subcomplex of X, then A is a closed subset of X and inherits a natural CW complex structure.

2.19   THEOREM   If A is a subcomplex of a finite CW complex X, then A is a strong deformation retract of some compact neighborhood of A in X.

PROOF   Denote by N the number of cells in $X - A$. We proceed by induction on N. If $N = 0$ the result is trivial and if $N = 1$ we may adapt the proof of Lemma 2.17 to give the desired result.

So suppose the result is true for any finite CW pair (Y, B) where the number of cells in $Y - B$ is $N - 1$. Let E_i^m be a cell of maximal dimension in $X - A$, and define $X_1 = X - E_i^m$. Note that X_1 must be a finite CW complex since any cell in $X - E_i^m$ either lies in A so that its boundary must also lie in A or has dimension less than or equal to m. In either case its boundary does not meet E_i^m. Moreover, A is a subcomplex of X_1.

Now the number of cells in $X_1 - A$ is $N - 1$, so by the inductive hypothesis there exists a compact neighborhood U_1 of A in X_1 such that A is a strong deformation retract of U_1.

There is a relative homeomorphism

$$\phi: \ (D^m, S^{m-1}) \rightarrow (\overline{E_i^m}, \overline{E_i^m} - E_i^m)$$

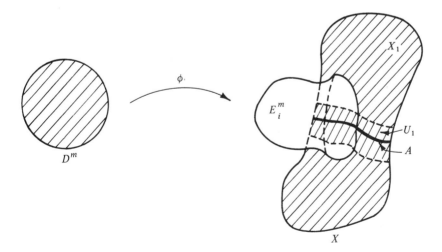

Figure 2.9

given by the structure of X as a finite CW complex (Figure 2.9). Define the radial projection map

$$r: \quad D^m - \{0\} \rightarrow S^{m-1}$$

by $r(x) = x/\| x \|$.

Since U_1 is a compact subset of X_1, $\phi^{-1}(U_1)$ is a compact subset of S^{m-1}. Now define

$$V = \{\phi(x) \mid \quad x \in D^m, \ \| x \| \geq \tfrac{1}{2} \text{ and } r(x) \in \phi^{-1}(U_1)\}.$$

Then V is a compact subset of X which admits a strong deformation retraction onto $U_1 \cap V$. Thus, $V \cup U_1$ is a compact subset of X which admits a strong deformation retraction onto A.

We must now make certain that the interior of $V \cup U_1$ contains A. Let y be in the interior of U_1 in X_1. If y is not in V, y must also be in the interior of U_1 in X, hence also in the interior of $V \cup U_1$. So suppose y is in V or, in other words, y is in $(\overline{E_i^m} - E_i^m)$. Now ϕ^{-1} of the interior of U_1 in X_1 is an open subset of S^{m-1} containing the compact set $\phi^{-1}(y)$. By the description of V it follows that $\phi^{-1}(y)$ is contained in the interior of $\phi^{-1}(V)$ in D^m. Thus, y must be in the interior of V in $\overline{E_i^m}$.

Therefore, we have shown that any point in the interior of U_1 in X_1 lies in the interior of $V \cup U_1$ in X. So $V \cup U_1$ gives the desired compact neighborhood of A in X.   □

Note that as an immediate consequence of this result, the conclusions in Theorem 2.14, Corollary 2.15 and the relative homeomorphism theorem, Theorem 2.16, will hold whenever the spaces involved are finite CW pairs. These have some very useful applications.

2.20   PROPOSITION   If X is a finite CW complex and X^k is the k-skeleton of X, then $H_j(X^k, X^{k-1}) = 0$ for $j \neq k$ and $H_k(X^k, X^{k-1})$ is a free abelian group with one basis element for each k-cell of X.

PROOF X^{k-1} is a subcomplex of X^k, so by Theorem 2.19 it is a strong deformation retract of a compact neighborhood in X^k. Since X is a finite CW complex, there is a relative homeomorphism

$$\phi: \quad (D_1^k \cup \cdots \cup D_r^k, S_1^{k-1} \cup \cdots \cup S_r^{k-1}) \rightarrow (X^k, X^{k-1}).$$

Then applying Theorem 2.16 yields the desired result from the corresponding fact about

$$H_*(D_1^k \cup \cdots \cup D_r^k, S_1^{k-1} \cup \cdots \cup S_r^{k-1}).$$ □

For any finite CW complex X define

$$C_k(X) = H_k(X^k, X^{k-1}).$$

Then $C_*(X) = \sum C_k(X)$ is a graded group which is nonzero in only finitely many dimensions, moreover it is free abelian and finitely generated in each dimension. The connecting homomorphism of the triple (X^k, X^{k-1}, X^{k-2}) defines an operator

$$\partial: \quad C_k(X) \to C_{k-1}(X).$$

Recall that these connecting homomorphisms may be factored in the following way:

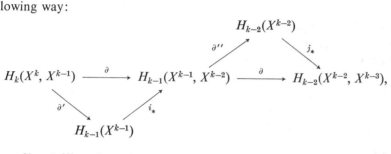

where ∂' and ∂'' are boundary operators for the respective pairs and i and j are inclusions of pairs. So $\partial \circ \partial = j_* \circ \partial'' \circ i_* \circ \partial'$. But $\partial'' \circ i_*$ is the composition of two consecutive homomorphisms in the exact sequence of the pair (X^{k-1}, X^{k-2}) and hence must be zero. Therefore, $\partial \circ \partial = 0$ and $\{C_*(X), \partial\}$ is a chain complex. Of course, the obvious question is then to ask how the homology of this chain complex is related to the singular homology of X.

2.21   THEOREM   If X is a finite CW complex, then

$$H_k(C_*(X)) \approx H_k(X) \qquad \text{for each} \quad k.$$

NOTE   This is an extreme simplification. The chain complex used in defining $H_*(X)$ was, in general, a free abelian group with an uncountable basis. Here we have reduced the chain complex, not only to a finite basis, but these generators are in one-to-one correspondence with the cells of X.

PROOF   We must analyze the composition

$$H_{k+1}(X^{k+1}, X^k) \xrightarrow{\partial_1} H_k(X^k, X^{k-1}) \xrightarrow{\partial_2} H_{k-1}(X^{k-1}, X^{k-2})$$

and show that kernel ∂_2/image $\partial_1 \approx H_k(X)$.

First consider the diagram

$$0$$
$$\downarrow$$
$$H_k(X^{k+1}, X^{k-2})$$
$$\downarrow j_*$$

$$H_{k+1}(X^{k+1}, X^k) \xrightarrow{\partial_1} H_k(X^k, X^{k-1}) \xrightarrow{i_*} H_k(X^{k+1}, X^{k-1}) \to 0$$

with ∂_2 (diagonal) and ∂_3 to

$$H_{k-1}(X^{k-1}, X^{k-2})$$

in which i_* and j_* are induced by inclusion maps of pairs. The row and the column are exact sequences of triples in which the zeros appear by Proposition 2.20. The triangle commutes by the naturality of the boundary operators.

Let $x \in$ kernel ∂_2. Then $\partial_3 i_*(x) = 0$ and $i_*(x) = j_*(y)$ for some $y \in H_k(X^{k+1}, X^{k-2})$. Note that since j_* is a monomorphism, this y is uniquely determined. Thus, we define a homomorphism

$$\phi: \text{ kernel } \partial_2 \to H_k(X^{k+1}, X^{k-2})$$

by $\phi(x) = y$.

If $y' \in H_k(X^{k+1}, X^{k-2})$ then $j_*(y')$ is in the image of i_* because i_* is an epimorphism. So there exists an $x' \in H_k(X^k, X^{k-1})$ with $i_*(x') = j_*(y')$. Then $\partial_2(x') = \partial_3 i_*(x') = \partial_3 j_*(y') = 0$ so x' is in the kernel of ∂_2 and $\phi(x') = y'$. We conclude that ϕ is an epimorphism.

Since $i_* \circ \partial_1 = 0$, it is apparent that the image of ∂_1 is contained in the kernel of ϕ. On the other hand, let $x \in$ kernel ∂_2 with $\phi(x) = 0$. But the fact that j_* is a monomorphism implies that $i_*(x) = 0$. Then by exactness x is in the image of ∂_1. Hence, we have shown that ϕ is an epimorphism with kernel given by the image of ∂_1 and we conclude that

$$\phi: \text{ ker } \partial_2/\text{im } \partial_1 \xrightarrow{\approx} H_k(X^{k+1}, X^{k-2}).$$

In the remainder of the proof we show that

$$H_k(X^{k+1}, X^{k-2}) \approx H_k(X).$$

Suppose that X is n-dimensional, so that

$$H_k(X) = H_k(X^n, X^{-1}).$$

Consider the sequence of homomorphisms

$$H_k(X) = H_k(X^n, X^{-1}) \to H_k(X^n, X^0) \to \cdots \to H_k(X^n, X^{k-2}),$$

each induced by an inclusion of pairs. In general, a homomorphism in this sequence is a part of the exact sequence of a triple

$$H_k(X^i, X^{i-1}) \to H_k(X^n, X^{i-1}) \to H_k(X^n, X^i) \xrightarrow{\ \varDelta\ } H_{k-1}(X^i, X^{i-1}),$$

where $i \le k - 2$. But by Proposition 2.20 for this range of values of i the first and last group must be zero. Hence, each homomorphism in the sequence is an isomorphism and

$$H_k(X) \approx H_k(X^n, X^{k-2}).$$

Similarly the homomorphisms

$$H_k(X^{k+1}, X^{k-2}) \to H_k(X^{k+2}, X^{k-2}) \to \cdots \to H_k(X^n, X^{k-2})$$

induced by inclusion maps are all isomorphism, so that

$$H_k(X^n, X^{k-2}) \approx H_k(X^{k+1}, X^{k-2})$$

and the proof is complete. $\square$

A map $f: X \to Y$ between finite CW complexes is *cellular* if $f(X^k) \subseteq Y^k$ for each integer k. Although we do not need this at the moment, it is true that any map $g: X \to Y$ is homotopic to a cellular map. For a proof of this result see Brown [1968, p. 261]. If $f: X \to Y$ is cellular, then f defines a map of pairs

$$f: \quad (X^k, X^{k-1}) \to (Y^k, Y^{k-1})$$

for each k, and hence a chain mapping

$$f_*: \quad C_*(X) \to C_*(Y).$$

One should check that the homomorphism induced by f_* on the homology of the chain complex $C_*(X)$ corresponds with the homomorphism induced by f on $H_*(X)$ under the isomorphism of Theorem 2.21.

We now want to compute the homology of $\mathbb{R}P(n)$. To do this, give S^n the structure of a finite CW complex so that the k-skeleton is S^k. That is

$$S^0 \subseteq S^1 \subseteq \cdots \subseteq S^k \subseteq \cdots \subseteq S^n$$

so that there are two cells in each dimension, denoted by E_+^k and E_-^k. Similarly give $RP(n)$ the structure of a finite CW complex so that $RP(k)$ is the k-skeleton. Thus

$$RP(0) \subseteq RP(1) \subseteq \cdots \subseteq RP(k) \subseteq \cdots \subseteq RP(n)$$

and there is one cell in each dimension.

   With these structures, the identification map $\pi: S^n \to RP(n)$ is cellular. By Proposition 2.20 the group $C_k(RP(n))$ is infinite cyclic for $0 \le k \le n$ and we denote a generator by e_k'. In order to compute the homology of $RP(n)$ we need to know what the boundary operator $\partial: C_k(RP(n)) \to C_{k-1}(RP(n))$ does to the element e_k'.

   To answer this question we first study the situation in S^n. Recall that the antipodal map of S^n, $A: S^n \to S^n$, is cellular and furthermore maps E_+^k homeomorphically onto E_-^k and vice versa for each k. Denote by F^k the composition of maps of pairs

$$(D^k, S^{k-1}) \xrightarrow{\approx} (\bar{E}_+^k, S^{k-1}) \xrightarrow{\text{incl}} (S^k, S^{k-1}).$$

If we choose a generator i_k of $H_k(D^k, S^{k-1})$, then $F_*^k(i_k) = e_k$ is a basis element in $H_k(S^k, S^{k-1}) = C_k(S^n)$. We view e_k as the basis element corresponding to the cell E_+^k. Since the following diagram commutes

$$
\begin{array}{ccc}
(D^k, S^{k-1}) \xrightarrow{\approx} (\bar{E}_+^k, S^{k-1}) & \xrightarrow{\text{incl}} & (S^k, S^{k-1}) \\
\Big\downarrow A & & \Big\downarrow A \\
(\bar{E}_-^k, S^{k-1}) \xrightarrow{\text{incl}} & (S^k, S^{k-1})
\end{array}
$$

we may take the element $A_*(e_k)$ to be the basis element corresponding to the cell E_-^k. Thus, $C_k(S^n)$ is the free abelian group with basis $\{e_k, A_*(e_k)\}$.

   To determine the boundary operator $\partial: C_k(S^n) \to C_{k-1}(S^n)$ consider the following diagram:

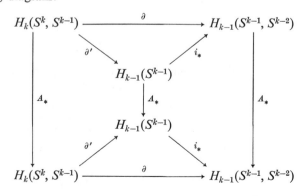

in which each triangle and rectangle is commutative. The homomorphism A_* in the center has been previously computed, specifically it is multiplication by $(-1)^k$. Starting with $e_k \in H_k(S^k, S^{k-1}) = C_k(S^n)$ we have

$$\partial A_*(e_k) = i_* \partial' A_*(e_k) = i_* A_* \partial'(e_k)$$
$$= (-1)^k i_* \partial'(e_k) = (-1)^k \partial(e_k).$$

Thus, $e_k + (-1)^{k+1} A_*(e_k)$ is a cycle in $C_k(S^n)$.

In fact, the set of cycles in $C_k(S^n)$ is an infinite cyclic subgroup generated by $e_k + (-1)^{k+1} A_*(e_k)$. Before proceeding with the proof, note that this algebraic fact is entirely reasonable from a geometric viewpoint. Since e_k and $A_*(e_k)$ correspond to the upper and lower halves of the sphere S^k, and they are being combined in such a way that the respective boundaries will cancel each other, geometrically we see this generating cycle as the sphere S^k itself. So suppose

$$0 = \partial(n_1 e_k + n_2 A_*(e_k))$$
$$= n_1 \partial e_k + n_2 \partial A_* e_k$$
$$= n_1 \partial e_k + (-1)^k n_2 \partial e_k$$
$$= (n_1 + (-1)^k n_2) \partial e_k.$$

Since $C_{k-1}(S^n)$ is free abelian, it must be true that either $\partial e_k = 0$ or $(n_1 + (-1)^k n_2) = 0$. Suppose $\partial e_k = 0$. Then also $\partial A_* e_k = 0$ and $\partial : C_k(S^n) \to C_{k-1}(S^n)$ is identically zero. But we have observed that there are non-trivial cycles in $C_{k-1}(S^n)$, so if $k > 1$, these cycles must bound because $H_{k-1}(S^n) = 0$ in this range. [It is also easy to see that $\partial : C_1(S^n) \to C_0(S^n)$ cannot be identically zero because every element of $C_0(S^n)$ is a cycle.] This contradiction implies that $\partial e_k \neq 0$ and we conclude that $n_1 + (-1)^k n_2 = 0$ or $n_2 = (-1)^{k+1} n_1$. Therefore

$$n_1 e_k + n_2 A_* e_k = n_1(e_k + (-1)^{k+1} A_* e_k)$$

as desired.

Since $H_k(S^n) = 0$ for $0 < k < n$, we must have

$$\partial(e_{k+1}) = \pm(e_k + (-1)^{k+1} A_*(e_k)).$$

Once again, this formula may be shown to hold as well for $k = 0$. We may as well suppose the sign is $+$.

The identification map $\pi : (S^k, S^{k-1}) \to (\mathbb{R}P(k), \mathbb{R}P(k-1))$ is a relative homeomorphism on the closure of each k-cell. The generator e_k' could have been chosen so that $e_k' = \pi_*(e_k)$. Then

$$\pi_*(A_*e_k) = (\pi \circ A)_*(e_k) = \pi_*(e_k) = e_k'.$$

Therefore, the boundary operator in the chain complex $C_*(\mathbb{R}P(n))$ is given by

$$\begin{aligned}
\partial(e_{k+1}') = \partial \pi_*(e_{k+1}) &= \pi_* \partial(e_{k+1}) \\
&= \pi_*(e_k + (-1)^{k+1} A_* e_k) \\
&= e_k' + (-1)^{k+1} e_k' \\
&= \begin{cases} 2e_k' & \text{for } k \text{ odd} \\ 0 & \text{for } k \text{ even.} \end{cases}
\end{aligned}$$

This completely determines the boundary operator in the chain complex $C_*(\mathbb{R}P(n))$ so that we may apply Theorem 2.21 to conclude the following:

2.22   PROPOSITION   The homology groups of real projective space are given by

$$H_i(\mathbb{R}P(n)) \approx \begin{cases} Z & \text{for } i = 0 \\ Z_2 & \text{for } i \text{ odd,} \quad 0 < i < n \\ Z & \text{for } i \text{ odd,} \quad i = n \\ 0 & \text{otherwise.} \end{cases} \quad \square$$

Recall that the *rank* of a finite generated abelian group A is given by

$$\text{rank } A = \text{lub}\{n \mid \text{there exists a free abelian subgroup } B \subseteq A \text{ with}$$

basis having exactly n elements$\}$.

If A and B are isomorphic abelian groups, then rank $A =$ rank B. If H is a subgroup of a finitely generated abelian group G, then

$$\text{rank } G/H = \text{rank } G - \text{rank } H.$$

2.23   PROPOSITION   If (X, A) is a finite CW pair, then $H_*(X, A)$ is a finitely generated abelian group.

PROOF   By Corollary 2.15 we know that $H_*(X, A) \approx \tilde{H}_*(X/A)$, and since $H_*(X/A) \approx \tilde{H}_*(X/A) \oplus Z$, it is sufficient to show that $H_*(X/A)$ is finitely generated. X/A may be given the structure of a finite CW complex directly from the structure of X and A. The cells of X/A correspond to the cells of X which are not in A together with one 0-cell corresponding to A, thus dim $(X/A) \le$ dim X. It follows from Theorem 2.21 that $H_k(X/A)$ is the quotient of a finitely generated abelian group by a subgroup, and is nonzero for only finitely many values of k. Therefore, $H_*(X/A)$ is finitely generated and the result follows. □

For a space X the ith *Betti number* of X, $b_i(X)$, is the rank of $H_i(X)$. From Proposition 2.23 we see that if X is a finite CW complex, $b_i(X)$ is finite for all i, and nonzero for only finitely many values of i. It was noted previously that $b_0(X)$ is the number of path components in X. In a corresponding sense the number $b_i(X)$ is a measure of a form of higher-dimensional connectivity of X. The *Euler characteristic* of X is given by

$$\chi(X) = \sum_i (-1)^i b_i(X).$$

2.24 PROPOSITION   If X is a finite CW complex with α_i cells in dimension i, then

$$\sum (-1)^i \alpha_i = \chi(X).$$

PROOF   Exercise 7. □

Exercise 8 If X and Y are finite CW complexes, show that

$$\chi(X \times Y) = \chi(X) \cdot \chi(Y).$$

chapter 3

THE EILENBERG–STEENROD
AXIOMS

Following the necessary algebraic preliminaries, we introduce the homology of a space with coefficients in an arbitrary abelian group. Combined with the results of the previous chapters this establishes the existence of homology theories satisfying the Eilenberg–Steenrod axioms for arbitrary coefficient groups. The corresponding uniqueness theorem is proved in the category of finite CW complexes. Finally, the singular cohomology groups are introduced and shown to satisfy the contravariant analogs of the axioms.

If A, B, and C are abelian groups, a mapping

$$\phi: \quad A \times B \to C$$

is *bilinear* (or is a *bihomomorphism*) if

$$\phi(a_1 + a_2, b) = \phi(a_1, b) + \phi(a_2, b)$$

and

$$\phi(a, b_1 + b_2) = \phi(a, b_1) + \phi(a, b_2).$$

Note that if $A \times B$ is given the usual product group structure, ϕ will not be a homomorphism except in very special cases.

Denote by $F(A \times B)$ the free abelian group generated by $A \times B$. An element of $F(A \times B)$ has the form

$$\Sigma\, n_i(a_i, b_i),$$

where the sum is finite, $a_i \in A$, $b_i \in B$ and n_i is an integer. Let $R(A \times B)$ be the subgroup of $F(A \times B)$ generated by elements of the form

$$(a_1 + a_2, b) - (a_1, b) - (a_2, b)$$

or

$$(a, b_1 + b_2) - (a, b_1) - (a, b_2),$$

where $a, a_1, a_2 \in A$ and $b, b_1, b_2 \in B$. Then the *tensor product* of A and B is defined to be

$$A \otimes B = F(A \times B)/R(A \times B).$$

Note that if $\phi: A \times B \to C$ is any function, there exists a unique extension of ϕ to a homomorphism

$$\phi': \quad F(A \times B) \to C.$$

Moreover, if ϕ is a bihomomorphism, then ϕ' is zero on the subgroup $R(A \times B)$, so that there is induced a homomorphism

$$\phi'': \quad A \otimes B \to C,$$

which is uniquely determined by ϕ.

This universal property with respect to bilinear maps can be used to characterize the tensor product. There exists a bilinear map

$$\tau: \quad A \times B \to A \otimes B$$

defined by taking (a, b) into $a \otimes b$, the coset containing (a, b). Given a bihomomorphism $\phi: A \times B \to C$ we have seen that there exists a unique homomorphism $\phi'': A \otimes B \to C$ such that commutativity holds in

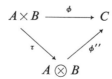

On the other hand, if G is abelian and $\tau': A \times B \to G$ is a bihomomorphism

whose image generates G, such that any bihomomorphism $\phi\colon A \times B \to C$ can be lifted through G, then G is isomorphic to $A \otimes B$.

Since the elements (a, b) generate $F(A \times B)$, it follows that the elements $a \otimes b$ generate $A \otimes B$. Note that in $A \otimes B$ we have

$$n(a \otimes b) = (na) \otimes b = a \otimes (nb) \qquad \text{for any integer} \quad n;$$

$$0 \otimes b = 0 = a \otimes 0 \qquad \text{for all} \quad a \text{ and } b;$$

$$(a_1 + a_2) \otimes (b_1 + b_2) = (a_1 + a_2) \otimes b_1 + (a_1 + a_2) \otimes b_2$$
$$= a_1 \otimes b_1 + a_2 \otimes b_1 + a_1 \otimes b_2 + a_2 \otimes b_2.$$

3.1 PROPOSITION   There is a unique isomorphism

$$\theta\colon \quad A \otimes B \approx B \otimes A$$

such that $\theta(a \otimes b) = (b \otimes a)$.

PROOF   Define $\mu\colon A \times B \to B \otimes A$ by $\mu(a, b) = b \otimes a$. This is a well-defined bihomomorphism. Thus, there exists a unique homomorphism $\theta\colon A \otimes B \to B \otimes A$ with $\theta(a \otimes b) = b \otimes a$. Similarly there exists a homomorphism $\theta'\colon B \otimes A \to A \otimes B$ such that $\theta'(b \otimes a) = a \otimes b$. Then the compositions $\theta \circ \theta'$ and $\theta' \circ \theta$ are the identity on respective generating sets, and it follows that θ is an isomorphism with inverse θ'. $\square$

3.2 PROPOSITION   Given homomorphisms $f\colon A \to A'$ and $g\colon B \to B'$, there exists a unique homomorphism

$$f \otimes g\colon \quad A \otimes B \to A' \otimes B'$$

with

$$f \otimes g(a \otimes b) = f(a) \otimes g(b).$$

PROOF   Define a mapping $\mu\colon A \times B \to A' \otimes B'$ by $\mu(a, b) = f(a) \otimes g(b)$, and observe that μ is well defined and bilinear. Thus, there exists a unique homomorphism $\theta\colon A \otimes B \to A' \otimes B'$ with $\theta(\tau(a, b)) = \mu(a, b)$ or $\theta(a \otimes b) = f(a) \otimes g(b)$. This θ is the desired $f \otimes g$. $\square$

3.3 PROPOSITIONS   (a) If $f\colon A \to A'$, $f'\colon A' \to A''$ and $g\colon B \to B'$, $g'\colon B' \to B''$, then

$$(f' \circ f) \otimes (g' \circ g) = (f' \otimes g') \circ (f \otimes g);$$

(b)   if $A \approx \sum A_j$, then $A \otimes B \approx \sum (A_j \otimes B)$;

(c)   if for each j in some index set J there is a homomorphism $f_j \colon A \to A'$ such that for any $a \in A$, $f_j(a)$ is nonzero for only finitely many values of j, then we can define $\sum f_j \colon A \to A'$. For any homomorphism $g \colon B \to B'$ it follows that

$$(\textstyle\sum f_j) \otimes g = \sum (f_j \otimes g);$$

(d)   for any abelian group A, $Z \otimes A \approx A$;

(e)   if A is a free abelian group with basis $\{a_i\}$ and B is a free abelian group with basis $\{b_j\}$, then $A \otimes B$ is a free abelian group with basis $\{a_i \otimes b_j\}$.

PROOF   We prove only Part (d). Note that Part (e) follows from Parts (d) and (b) and Proposition 3.1.

Define $\mu \colon Z \times A \to A$ by $\mu(n, a) = na$. Then μ is bilinear, so there exists a unique lifting $\theta \colon Z \otimes A \to A$ with $\theta(n \otimes a) = n \cdot a$. Now define $\theta' \colon A \to Z \otimes A$ by $\theta'(a) = 1 \otimes a$ and observe that

$$\theta\theta'(a) = \theta(1 \otimes a) = a$$

and

$$\theta'\theta(n \otimes a) = \theta'(na) = 1 \otimes na = n \otimes a.$$

Thus, θ and θ' behave as inverses on generating sets and θ is an isomorphism. $\square$

Now suppose that A' and B' are subgroups of A and B, respectively. We want to describe the tensor product $A/A' \otimes B/B'$. Denote by $\pi_1 \colon A \to A/A'$ and $\pi_2 \colon B \to B/B'$ the quotient homomorphisms. Then by Proposition 3.2 there is a homomorphism

$$\pi_1 \otimes \pi_2 \colon \quad A \otimes B \to A/A' \otimes B/B'.$$

If $a' \in A'$ and $b \in B$, then $\pi_1 \otimes \pi_2(a' \otimes b) = \pi_1(a') \otimes \pi_2(b) = 0$. Similarly if $a \in A$ and $b' \in B'$, $\pi_1 \otimes \pi_2(a \otimes b') = 0$. Thus, if we denote by

$$i_1 \colon \quad A' \to A \qquad \text{and} \qquad i_2 \colon \quad B' \to B,$$

the inclusion homomorphisms

$$H = \operatorname{im}(i_1 \otimes \operatorname{id}) + \operatorname{im}(\operatorname{id} \otimes i_2) \subseteq \ker \pi_1 \otimes \pi_2.$$

This means that $\pi_1 \otimes \pi_2$ induces a homomorphism

$$\Phi \colon \quad A \otimes B/H \to A/A' \otimes B/B'.$$

We now want to show that Φ is an isomorphism. Define a function

$$\Psi: \quad A/A' \times B/B' \to A \otimes B/H$$

by $\Psi(\{a\}, \{b\}) = \{a \otimes b\}$, where $\{\ \}$ denotes the respective coset. It is evident that this is well defined, since if $a' \in A'$, then

$$\Psi(\{a'\}, \{b\}) = \{a' \otimes b\} = 0,$$

and similarly for $b' \in B'$. Ψ is also bilinear, so there exists a unique homomorphism

$$\theta: \quad A/A' \otimes B/B' \to A \otimes B/H.$$

The homomorphisms Φ and θ are easily seen to be inverses of each other, so we have proved the following.

3.4   PROPOSITION   If $i_1: A' \to A$ and $i_2: B' \to B$ are inclusions of subgroups, then

$$A/A' \otimes B/B' \approx \frac{A \otimes B}{\mathrm{im}(i_1 \otimes \mathrm{id}) + \mathrm{im}(\mathrm{id} \otimes i_2)}. \qquad \square$$

Example $Z_p \otimes Z_q \approx Z_{(p,q)}$, where (p, q) is the greatest common divisor of p and q. To see this, let $(p, q) = r$ so that $p = r \cdot s$, $q = r \cdot t$ with $(s, t) = 1$. Denote by $pZ \subseteq Z$ the subgroup divisible by p and identify $Z_p = Z/pZ$ and $Z_q = Z/qZ$.

Therefore

$$Z_p \otimes Z_q = Z/pZ \otimes Z/qZ$$

$$\approx \frac{Z \otimes Z}{\mathrm{im}(i_1 \otimes \mathrm{id}) + \mathrm{im}(\mathrm{id} \otimes i_2)}$$

$$\approx \frac{Z}{\mathrm{im}\, i_1 + \mathrm{im}\, i_2} \approx \frac{Z}{rsZ + rtZ} \approx Z/rZ$$

$$= Z_r = Z_{(p,q)}.$$

Specifically then

$$Z_2 \otimes Z_2 \approx Z_2, \quad Z_2 \otimes Z_3 = 0, \quad Z_6 \otimes Z_{15} \approx Z_3, \quad \text{and so forth.}$$

3.5   PROPOSITION   If $A \xrightarrow{\alpha} B \xrightarrow{\beta} C \to 0$ is an exact sequence, then for any abelian group D

$$A \otimes D \xrightarrow{\alpha \otimes \mathrm{id}} B \otimes D \xrightarrow{\beta \otimes \mathrm{id}} C \otimes D \to 0$$

is exact.

so that $f \otimes$ id is also a chain map. Suppose T is a chain homotopy between chain maps f_0 and f_1, that is,

$$\partial' T + T\partial = f_1 - f_0.$$

Then

$$\begin{aligned}
(\partial' \otimes \text{id})(T \otimes \text{id}) + (T \otimes \text{id})(\partial \otimes \text{id}) &= \partial' T \otimes \text{id} + T\partial \otimes \text{id} \\
&= (\partial' T + T\partial) \otimes \text{id} \\
&= (f_1 - f_0) \otimes \text{id} \\
&= f_1 \otimes \text{id} - f_0 \otimes \text{id}.
\end{aligned}$$

Hence, $T \otimes$ id is a chain homotopy between the chain maps $f_1 \otimes$ id and $f_0 \otimes$ id.

Now fix the abelian group G. For each pair of spaces (X, A) we may use the free chain complex $S_*(X, A)$ to construct a new chain complex $S_*(X, A) \otimes G$. This chain complex, denoted $S_*(X, A; G)$, is the *singular chain complex of* (X, A) *with coefficients in* G. Since there is a natural isomorphism

$$S_*(X, A) \otimes Z \approx S_*(X, A),$$

we refer to $S_*(X, A)$ as the singular chain complex with *integral coefficients*. The homology of $S_*(X, A; G)$ is denoted by $H_*(X, A; G)$. Note that if $f: (X, A) \to (X', A')$ is a map of pairs, it follows from the preceding comments that there is an induced homomorphism

$$f_*: \quad H_*(X, A; G) \to H_*(X', A'; G).$$

In some applications it is desirable to have additional structures on these homology groups. For example suppose that R is an associative ring and G is a right R-module. Then $S_n(X, A; G)$ may easily be given the structure of a right R-module in such a way that the boundary operators and induced homomorphisms are all homomorphisms of R-modules. In particular, if R is a field, then each $H_n(X, A; G)$ is a vector space over R. Note that for any R we know that R is a free module over itself, so that $S_n(X, A; R)$ is the free R-module generated by the singular n-simplices of X mod A.

Suppose that (X, A, B) is a triple of spaces. We have observed previously that there is a short exact sequence of chain maps

$$0 \to S_*(A, B) \to S_*(X, B) \to S_*(X, A) \to 0.$$

Since each chain complex is free, it follows from Exercises 1 and 2 that the exactness is preserved when we tensor throughout with G. Thus

$$0 \to S_*(A, B; G) \to S_*(X, B; G) \to S_*(X, A; G) \to 0$$

is a short exact sequence of chain complexes and chain maps. There results the long exact sequence of the triple (X, A, B) for homology with coefficients in G.

As in the case of integral coefficients it is easy to show that

$$H_n(\text{pt}; G) \approx \begin{cases} G & \text{for} \quad n = 0 \\ 0 & \text{otherwise.} \end{cases}$$

Returning to the general case of a free chain complex C, we now consider the problem of relating the homology of $C \otimes G$ to the homology of C. For example, suppose that $x \in C_n$ such that px is a boundary for some integer p. Since C is free, this implies that x must be a cycle, so x represents a homology class of order p. Let $b \in C_{n+1}$ with $\partial b = px$. Note that b is not a cycle unless $x = 0$. If we now tensor C with $G = Z_p$, the element $b \otimes 1$ in $C_{n+1} \otimes Z_p$ has

$$(\partial \otimes \text{id})(b \otimes 1) = \partial b \otimes 1 = px \otimes 1 = x \otimes p = 0.$$

Thus, $b \otimes 1$ is a cycle in $C_{n+1} \otimes Z_p$, where b had not been a cycle previously. In this way we see how torsion common to $H_n(C)$ and G produces new homology classes in $H_{n+1}(C \otimes G)$.

Before proceeding we note the easily proved algebraic fact that if $f: G \to G'$ and $g: G' \to G$ are homomorphisms of abelian groups with $g \circ f = \text{identity}$, then

$$G' = \text{im } f \oplus \text{ker } g.$$

As usual we denote by $B_n \subseteq Z_n \subseteq C_n$ the subgroups of boundaries and cycles, respectively. If C is a free chain complex, then each B_n and Z_n will be a free abelian group. Fix an abelian group G and consider the short exact sequence

$$0 \to Z_n \to C_n \xrightarrow[\substack{\partial \\ \longleftarrow \\ \gamma}]{} B_{n-1} \to 0.$$

First note that since B_{n-1} is free, the sequence splits. That is, if $\{x_i\}$ is a basis for B_{n-1}, for each i there exists an element $c_i \in C_n$ with $\partial c_i = x_i$. Define $\gamma(x_i) = c_i$ and note that γ extends uniquely to a homomorphism which splits the sequence.

Now since B_{n-1} is free, $\mathrm{Tor}(B_{n-1}, G) = 0$ and the short exact sequence is preserved when tensored with G,

$$0 \to Z_n \otimes G \to C_n \otimes G \underset{\gamma \,\otimes\, \mathrm{id}}{\overset{\partial \,\otimes\, \mathrm{id}}{\rightleftarrows}} B_{n-1} \otimes G \to 0.$$

This sequence is also split by the homomorphism $\gamma \otimes \mathrm{id}$.

On the other hand, the short exact sequence

$$0 \to B_n \overset{j}{\to} Z_n \to H_n(C) \to 0$$

is a free resolution of $H_n(C)$; hence, it yields the exact sequence

$$0 \to \mathrm{Tor}(H_n(C), G) \overset{g}{\to} B_n \otimes G \overset{j \,\otimes\, \mathrm{id}}{\longrightarrow} Z_n \otimes G \to H_n(C) \otimes G \to 0.$$

We want to compute the homology of $C \otimes G$, given by kernel $\partial \otimes \mathrm{id}/$ image $\partial_1 \otimes \mathrm{id}$ in the following diagram:

$$
\begin{array}{ccc}
B_n \otimes G & \xrightarrow{\ j \,\otimes\, \mathrm{id}\ } & Z_n \otimes G \\
{\scriptstyle \partial_3 \,\otimes\, \mathrm{id}} \uparrow & & \downarrow {\scriptstyle k \,\otimes\, \mathrm{id}} \\
C_{n+1} \otimes G \xrightarrow{\ \partial_1 \,\otimes\, \mathrm{id}\ } & C_n \otimes G & \xrightarrow{\ \partial \,\otimes\, \mathrm{id}\ } C_{n-1} \otimes G \\
& {\scriptstyle \partial_2 \,\otimes\, \mathrm{id}} \searrow \quad \nearrow {\scriptstyle i \,\otimes\, \mathrm{id}} \\
& Z_{n-1} \otimes G &
\end{array}
$$

Note that by the above remarks both $i \otimes \mathrm{id}$ and $k \otimes \mathrm{id}$ are monomorphisms and $\partial_3 \otimes \mathrm{id}$ is an epimorphism. Thus, kernel $\partial \otimes \mathrm{id} =$ kernel $\partial_2 \otimes \mathrm{id}$ and image $\partial_1 \otimes \mathrm{id}$ may be identified with image $j \otimes \mathrm{id}$.

Now consider the groups and homomorphisms

$$0 \to \mathrm{Tor}(H_{n-1}(C), G) \overset{g}{\to} B_{n-1} \otimes G \overset{j \,\otimes\, \mathrm{id}}{\longrightarrow} Z_{n-1} \otimes G \to H_{n-1}(C) \otimes G \to 0$$

$$\partial_3 \otimes \mathrm{id} \updownarrow \gamma \otimes \mathrm{id}$$

$$C_n \otimes G$$

where the horizontal row is exact. As we have observed, the cycle group in $C_n \otimes G$ is the kernel of $(j \otimes \mathrm{id}) \circ (\partial_3 \otimes \mathrm{id})$. Since $\partial_3 \otimes \mathrm{id}$ is an epimorphism and kernel $j \otimes \mathrm{id} =$ image g, we have

$$\ker(j \otimes \mathrm{id}) \circ (\partial_3 \otimes \mathrm{id}) = (\partial_3 \otimes \mathrm{id})^{-1}(g(\mathrm{Tor}(H_{n-1}(C), G))).$$

Thus, there are homomorphisms

$$(\partial_3 \otimes \text{id})^{-1}g(\text{Tor}(H_{n-1}(C), G)) \underset{\gamma \otimes \text{id}}{\overset{\partial_3 \otimes \text{id}}{\rightleftarrows}} g(\text{Tor}(H_{n-1}(C), G))$$

for which the composition $(\partial_3 \otimes \text{id}) \circ (\gamma \otimes \text{id})$ is the identity. Combining these observations we have the cycle group expressed as a direct sum

$$\ker(j \otimes \text{id}) \circ (\partial_3 \otimes \text{id}) = \ker(\partial_3 \otimes \text{id}) \oplus (\gamma \otimes \text{id})g(\text{Tor}(H_{n-1}(C), G)).$$

Note that the first direct summand may be identified with $Z_n \otimes G$, while in the second, both g and $\gamma \otimes \text{id}$ are monomorphisms. Thus, we may identify the groups of cycles in $C_n \otimes G$ with the direct sum

$$Z_n \otimes G \oplus \text{Tor}(H_{n-1}(C), G).$$

Furthermore, the group of boundaries, which has been identified with the image of

$$B_n \otimes G \to Z_n \otimes G,$$

is contained entirely in the first summand.

Finally, recalling the exactness of

$$B_n \otimes G \to Z_n \otimes G \to H_n(C) \otimes G \to 0,$$

we conclude that the homology of the chain complex $C \otimes G$ is given by $H_n(C \otimes G) \approx H_n(C) \otimes G \oplus \text{Tor}(H_{n-1}(C), G)$. This completes the proof of the *universal coefficient theorem*:

3.6 THEOREM   If C is a free chain complex and G is an abelian group, then

$$H_n(C \otimes G) \approx H_n(C) \otimes G \oplus \text{Tor}(H_{n-1}(C), G). \qquad \square$$

3.7 COROLLARY   For any pair of spaces (X, A),

$$H_n(X, A; G) \approx H_n(X, A) \otimes G \oplus \text{Tor}(H_{n-1}(X, A), G). \qquad \square$$

Example   Recall that the integral homology groups of real projective space are given by

$$H_k(\mathbb{R}P(n)) \approx \begin{cases} Z & \text{for } k = 0 \text{ or for } k \text{ odd and } = n \\ Z_2 & \text{for } k \text{ odd, } 0 < k < n \\ 0 & \text{otherwise.} \end{cases}$$

Applying Corollary 3.7 to compute $H_*(\mathbb{R}P(n); Z_2)$ we first note that $\text{Tor}(Z_2, Z_2) \approx Z_2$ and $\text{Tor}(Z, Z_2) = 0$ were results from an earlier exercise. We are thus able to conclude that

$$H_k(\mathbb{R}P(n); Z_2) \approx \begin{cases} Z_2 & \text{for } 0 \leq k \leq n \\ 0 & \text{otherwise,} \end{cases}$$

where $H_k(\mathbb{R}P(n); Z_2)$ results from $H_k(\mathbb{R}P(n)) \otimes Z_2$ for $k = 0$ or for k odd, $0 < k \leq n$, and $H_k(\mathbb{R}P(n); Z_2)$ results from two-torsion in $H_{k-1}(\mathbb{R}P(n))$ for k even $0 < k \leq n$.

We are now in a position to characterize singular homology in terms of a set of axioms. Each of these axioms has been established previously as an intrinsic property of singular homology theory. Our main purpose here is to show that when restricted to a suitable category of spaces and maps, these axioms uniquely determine a homology theory. The formulation of the axioms and the proof of the uniqueness are due to Eilenberg and Steenrod [1952].

Suppose $\mathfrak{IC}$ is a function assigning to each pair of spaces (X, A) and integer n an abelian group $\mathfrak{IC}_n(X, A)$, and to each map of pairs $f \colon (X, A) \to (Y, B)$ a homomorphism $f_* \colon \mathfrak{IC}_n(X, A) \to \mathfrak{IC}_n(Y, B)$. Suppose further that for each n there is a homomorphism $\partial \colon \mathfrak{IC}_n(X, A) \to \mathfrak{IC}_{n-1}(A)$. This operation gives a *homology theory* if the following axioms are satisfied:

(1) if id: $(X, A) \to (X, A)$ is the identity map, then

$$\text{id}_* \colon \quad \mathfrak{IC}_n(X, A) \to \mathfrak{IC}_n(X, A)$$

is the identity homomorphism;

(2) if $f \colon (X, A) \to (X', A')$, $g \colon (X', A') \to (X'', A'')$ are maps of pairs, then

$$(g \circ f)_* = g_* \circ f_*;$$

(3) if $f \colon (X, A) \to (Y, B)$ is a map of pairs, then

$$\partial \circ f_* = (f|_A)_* \circ \partial;$$

(4) if $i \colon (A, \varnothing) \to (X, \varnothing)$ and $j \colon (X, \varnothing) \to (X, A)$ are inclusion maps, then the following sequence is exact:

$$\cdots \xrightarrow{\partial} \mathfrak{IC}_n(A) \xrightarrow{i_*} \mathfrak{IC}_n(X) \xrightarrow{j_*} \mathfrak{IC}_n(X, A) \xrightarrow{\partial} \mathfrak{IC}_{n-1}(A) \to \cdots;$$

(5) if $f, g \colon (X, A) \to (Y, B)$ are homotopic as maps of pairs, then $f_* = g_*$ as homomorphisms;

(6) if (X, A) is a pair and $U \subseteq A$ has $\bar{U} \subseteq \text{Int } A$, then the inclusion map
$i: (X-U, A-U) \to (X, A)$ has $i_*: \mathfrak{IC}_n(X-U, A-U) \to \mathfrak{IC}_n(X, A)$
an isomorphism;

(7) $\mathfrak{IC}_n(\text{pt}) = 0$ for $n \neq 0$ [$\mathfrak{IC}_0(\text{pt})$ is called the *coefficient group*].

3.8 THEOREM (*Existence*)   Given any abelian group G, there exists a homology theory with coefficient group G.

PROOF   Let $\mathfrak{IC}_n(X, A) = H_n(X, A; G)$, singular homology with coefficients in G. Then each of the axioms has been proved previously. $\square$

3.9 THEOREM (*Uniqueness*)   On the category of finite CW pairs and maps of pairs, homology theories are determined, up to isomorphism, by their coefficient groups. That is, if $\mathfrak{IC}_*$ and $\mathfrak{IC}_*'$ are homology theories and $h: \mathfrak{IC}_* \to \mathfrak{IC}_*'$ is a natural transformation (that is, it commutes with induced homomorphisms and boundary operators) such that $h: \mathfrak{IC}_0(\text{pt}) \to \mathfrak{IC}_0'(\text{pt})$ is an isomorphism, then $h: \mathfrak{IC}_n(X, A) \to \mathfrak{IC}_n'(X, A)$ is an isomorphism for each integer n and each finite CW pair (X, A).

PROOF   First note that the proofs of Theorems 2.14 and 2.16 (the relative homeomorphism theorem) only require that singular homology theory satisfy these axioms. So the analogs of these results will hold for any homology theory.

Denote the zero-sphere S^0 as the union of two points $S^0 = x \cup y$, and consider the diagram

$$\begin{array}{ccc} \mathfrak{IC}_k(y) & \xrightarrow{\;i_*\;} & \mathfrak{IC}_k(x \cup y, x) \\ \downarrow{\scriptstyle h} & & \downarrow{\scriptstyle h} \\ \mathfrak{IC}_k'(y) & \xrightarrow{\;i_*\;} & \mathfrak{IC}_k'(x \cup y, x) \end{array}$$

which commutes by the naturality of h. The horizontal maps are excision maps, so both horizontal homomorphisms are isomorphisms by Axiom 6. Since the first vertical homomorphism is an isomorphism by the hypothesis, we conclude that

$$h: \quad \mathfrak{IC}_k(S^0, x) \to \mathfrak{IC}_k'(S^0, x)$$

is an isomorphism for each k. Now consider the diagram

$$\begin{array}{ccccccccc} \mathfrak{IC}_{k+1}(S^0, x) & \xrightarrow{\;\partial\;} & \mathfrak{IC}_k(x) & \to & \mathfrak{IC}_k(S^0) & \to & \mathfrak{IC}_k(S^0, x) & \to & \mathfrak{IC}_{k-1}(x) \\ {\scriptstyle \approx}\downarrow{\scriptstyle h} & & {\scriptstyle \approx}\downarrow{\scriptstyle h} & & \downarrow{\scriptstyle h} & & {\scriptstyle \approx}\downarrow{\scriptstyle h} & & {\scriptstyle \approx}\downarrow{\scriptstyle h} \\ \mathfrak{IC}_{k+1}'(S^0, x) & \xrightarrow{\;\partial\;} & \mathfrak{IC}_k'(x) & \to & \mathfrak{IC}_k'(S^0) & \to & \mathfrak{IC}_k'(S^0, x) & \to & \mathfrak{IC}_{k-1}'(x) \end{array}$$

where the rows are exact by Axiom 4. By the five lemma (Exercise 4, Chapter 2), $h: \mathcal{H}_k(S^0) \to \mathcal{H}_{k'}'(S^0)$ is an isomorphism.

We now prove inductively that h is an isomorphism for spheres of all dimensions. Suppose $h: \mathcal{H}_k(S^{n-1}) \to \mathcal{H}_{k'}'(S^{n-1})$ is an isomorphism, $n > 0$. The n-disk D^n has the homotopy type of a point, so by using Axiom 5 we have an isomorphism $h: \mathcal{H}_k(D^n) \to \mathcal{H}_{k'}'(D^n)$. In the commutative diagram

$$
\begin{array}{ccc}
\mathcal{H}_k(D^n, S^{n-1}) & \xrightarrow{\pi_*} & \mathcal{H}_k(S^n, x) \\
\downarrow{h} & & \downarrow{h} \\
\mathcal{H}_{k'}'(D^n, S^{n-1}) & \xrightarrow{\pi_*} & \mathcal{H}_{k'}'(S^n, x)
\end{array}
$$

the horizontal homomorphisms are isomorphisms since they are induced by relative homeomorphisms, while the vertical homomorphism on the left is an isomorphism by the five lemma (Exercise 4, Chapter 2). So we conclude that $h: \mathcal{H}_k(S^n, x) \to \mathcal{H}_{k'}'(S^n, x)$ is an isomorphism, and again apply the five lemma to see that $h: \mathcal{H}_k(S^n) \to \mathcal{H}_{k'}'(S^n)$ is an isomorphism. This completes the inductive step.

We are now ready to prove the theorem by inducting on the number of cells in the finite CW complex, X. Of course the conclusion is true if X has only one cell, so suppose that h is an isomorphism for all complexes having less than m cells. Let X be a finite CW complex containing m cells. If dim $X = n$, pick a specific n-cell of X and denote by A the complement of this top dimensional cell. Then A is a subcomplex of X having $m - 1$ cells and there is a relative homeomorphism

$$\pi: \quad (D^n, S^{n-1}) \to (X, A).$$

In the commutative diagram

$$
\begin{array}{ccc}
\mathcal{H}_k(D^n, S^{n-1}) & \xrightarrow[\approx]{\pi_*} & \mathcal{H}_k(X, A) \\
\approx \downarrow{h} & & \downarrow{h} \\
\mathcal{H}_{k'}'(D^n, S^{n-1}) & \xrightarrow[\approx]{\pi_*} & \mathcal{H}_{k'}'(X, A)
\end{array}
$$

the horizontal homomorphisms are induced by relative homeomorphisms, so they are isomorphisms. The first vertical homomorphism is an isomorphism by the inductive argument above. Hence

$$h: \quad \mathcal{H}_k(X, A) \to \mathcal{H}_{k'}'(X, A)$$

is an isomorphism for each k. Finally, the five lemma together with the inductive hypothesis imply that

$$h: \quad \mathcal{H}_k(X) \to \mathcal{H}_k'(X)$$

is an isomorphism. This establishes the theorem for any finite CW complex, and the similar result for pairs follows by another application of the five lemma. $\square$

NOTE During recent years, many theories have been developed which satisfy all of the axioms except Axiom 7. These have been called "generalized homology theories" and include stable homotopy, various K-theories, and bordism theories. Some of these theories are able to detect invariants which cannot be detected by ordinary homology. As a result, problems have been solved by these techniques whose solutions in terms of singular homology were either extremely difficult or impossible. Certainly any thorough study of modern methods in algebraic topology should include a significant segment on generalized homology and cohomology theories.

We now want to introduce singular cohomology theory. If A and G are abelian groups, denote by $\text{Hom}(A, G)$ the abelian group of homomorphisms from A to G, where $(f + g)(a) = f(a) + g(a)$ for each a in A. If $\phi: A \to B$ is a homomorphism, there is an induced homomorphism

$$\phi^{\#}: \quad \text{Hom}(B, G) \to \text{Hom}(A, G)$$

defined by $\phi^{\#}(f) = f \circ \phi$. Note that if $\psi: B \to C$ is a homomorphism, then $(\psi \circ \phi)^{\#} = \phi^{\#} \circ \psi^{\#}$.

For a chain complex $\{C_n, \partial\}$ and an abelian group G, define abelian groups

$$C^n = \text{Hom}(C_n, G).$$

Then the boundary operator $\partial: C_{n+1} \to C_n$ has

$$\partial^{\#}: \quad C^n \to C^{n+1}$$

and the composition $\partial^{\#} \circ \partial^{\#} = (\partial \circ \partial)^{\#} = 0$. So this resembles a chain complex except that the indices are increased rather than decreased. This leads us to define a *cochain complex* to be a collection of abelian groups and homomorphisms $\{C^n, \delta\}$ where $\delta: C^n \to C^{n+1}$ and $\delta \circ \delta = 0$. The homomorphism δ is the *coboundary operator*.

Note that if $\{C^n, \delta\}$ is a cochain complex and we define $D_n = C^{-n}$ and $\partial = \delta: D_n \to D_{n-1}$, then $\{D_n, \partial\}$ becomes a chain complex. So the two notions are precisely dual to each other and the use of cochain complexes is mainly a convenience.

The basic definitions for chain complexes may be duplicated for cochain complexes. If $\{C^n, \delta\}$ and $\{D^n, \delta'\}$ are cochain complexes, a *cochain map f of degree k* is a collection of homomorphisms

$$f: \quad C^n \to D^{n+k}$$

such that $f \circ \delta = \delta' \circ f$. Two cochain maps f and g of degree zero are *cochain homotopic* if there is a collection of homomorphisms

$$T: \quad C^n \to D^{n-1}$$

such that $\delta'T + T\delta = f - g$. T is a *cochain homotopy*.

Note that if $\{C_n, \partial\}$ and $\{D_n, \partial'\}$ are chain complexes and $f, g: \{C_n, \partial\} \to \{D_n, \partial'\}$ are chain homotopic chain maps, then for any abelian group G the cochain maps

$$f^{\#}, g^{\#}: \quad \{\text{Hom}(D_n, G), \partial'^{\#}\} \to \{\text{Hom}(C_n, G), \partial^{\#}\}$$

are cochain homotopic.

Let $C = \{C^n, \delta\}$ be a cochain complex and define $Z^n(C) = \text{kernel}$ $\delta: C^n \to C^{n+1}$, the group of *n-cocycles*, and $B^n(C) = \text{image } \delta: C^{n-1} \to C^n$, the group of *n-coboundaries*. The *n*th *cohomology group* of C is then the quotient group

$$H^n(C) = Z^n(C)/B^n(C).$$

If $A = \{A^n\}$, $B = \{B^n\}$, and $C = \{C^n\}$ are cochain complexes and

$$0 \to A \to B \to C \to 0$$

is a short exact sequence of cochain maps of degree zero, then there exists a long exact sequence of cohomology groups

$$\cdots \to H^n(A) \to H^n(B) \to H^n(C) \xrightarrow{\Delta} H^{n+1}(A) \to \cdots,$$

where the connecting homomorphism Δ is defined in a fashion analogous to the connecting homomorphism for homology.

Now let (X, A) be a pair of spaces and G be an abelian group. Define

$$S^n(X, A; G) = \text{Hom}(S_n(X, A), G)$$

as the *n-dimensional cochain group of* (X, A) *with coefficients in G*. Let

$$\delta: \quad S^n(X, A; G) \to S^{n+1}(X, A; G)$$

be given by $\delta = \partial^{\#}$. This defines the singular cochain complex of (X, A) whose homology is the graded group

$$H^*(X, A; G)$$

as the *singular cohomology group of* (X, A) *with coefficients in G*. Each of the covariant properties of singular homology becomes a contravariant property of singular cohomology. In particular, if $f: (X, A) \to (Y, B)$ is a map of pairs, than there is induced a homomorphism

$$f^*: \quad H^*(Y, B; G) \to H^*(X, A; G).$$

If $g: (Y, B) \to (W, C)$ is another map of pairs, then $(gf)^* = f^* \circ g^*$. Let

$$0 \to F \xrightarrow{i} H \xrightarrow{\pi} K \to 0$$

be a short exact sequence of abelian groups and homomorphism which is split by a homomorphism $\gamma: H \to F$. If G is an abelian group and f is a nonzero element of $\operatorname{Hom}(K, G)$, then $\pi^{\#}(f) = f \circ \pi$ is a nonzero element of $\operatorname{Hom}(H, G)$ since π is an epimorphism. It is evident that $i^{\#} \circ \pi^{\#} = (\pi \circ i)^{\#} = 0$. On the other hand, let $h: H \to G$ be a homomorphism such that $i^{\#}(h) = h \circ i = 0$. Since h is zero on the image of $i = $ kernel of π, h may be factored through K. The resulting homomorphism $\bar{h}: K \to G$ will have $\pi^{\#}(\bar{h}) = h$. Thus, the kernel of $i^{\#}$ is equal to the image of $\pi^{\#}$.

Finally, since $\gamma \circ i$ is the identity on F, $(\gamma \circ i)^{\#}$ is the identity on $\operatorname{Hom}(F, G)$ But this implies that $i^{\#}$ is an epimorphism. Therefore, we have completed the proof of the following.

3.10 PROPOSITION If $0 \to F \xrightarrow{i} H \xrightarrow{\pi} K \to 0$ is a split exact sequence and G is an abelian group, then

$$0 \to \operatorname{Hom}(K, G) \xrightarrow{\pi^{\#}} \operatorname{Hom}(H, G) \xrightarrow{i^{\#}} \operatorname{Hom}(F, G) \to 0$$

is exact. $\square$

For example, if (X, A) is a pair of spaces, the sequence

$$0 \to S_*(A) \to S_*(X) \to S_*(X, A) \to 0$$

is split exact since $S_*(X, A)$ is a free chain complex. Thus, by Proposition 3.10

$$0 \to S^*(X, A; G) \to S^*(X; G) \to S^*(A; G) \to 0$$

is a short exact sequence of cochain complexes and cochain maps. By the previous remarks, this produces a long exact sequence in singular cohomology,

$$\cdots \to H^n(X, A; G) \to H^n(X; G) \to H^n(A; G) \to H^{n+1}(X, A; G) \to \cdots.$$

It is important to note the necessity of the hypothesis in Proposition 3.10 that the original sequence be split exact. For example, if $i: Z \to Z$ is the monomorphism given by $i(1) = 2$, then

$$i^\#: \quad \mathrm{Hom}(Z, Z_2) \to \mathrm{Hom}(Z, Z_2)$$

is zero and thus fails to be an epimorphism. The other conclusions of exactness will hold in general since they were established without using the fact that the sequence was split. As in the case of the tensor product, this failure to preserve short exact sequences may be measured.

Let E be an abelian group and take a free resolution of E,

$$0 \to R \xrightarrow{i} F \xrightarrow{\pi} E \to 0.$$

Then for any abelian group G the sequence

$$0 \to \mathrm{Hom}(E, G) \xrightarrow{\pi^\#} \mathrm{Hom}(F, G) \xrightarrow{i^\#} \mathrm{Hom}(R, G)$$

is exact by the proof of Proposition 3.10. Define

$$\mathrm{Ext}(E, G) = \mathrm{coker}\ i^\# = \mathrm{Hom}(R, G)/\mathrm{im}\ i^\#.$$

The basic properties of Ext are dual to those of Tor and may be established in the following exercises:

Exercise 3 (a) If E is free abelian, $\mathrm{Ext}(E, G) = 0$.

(b) $\mathrm{Ext}(E, G)$ is independent of the choice of the resolution for E.

(c) $\mathrm{Ext}(E, G)$ is contravariant in E and covariant in G; that is, given homomorphisms $f: E \to E'$ and $h: G \to G'$ there are induced homomorphisms $f^*: \mathrm{Ext}(E', G) \to \mathrm{Ext}(E, G)$ and $h_*: \mathrm{Ext}(E, G) \to \mathrm{Ext}(E, G')$.

(d) If $0 \xrightarrow{j} A \to B \xrightarrow{k} C \to 0$ is a short exact sequence, then exactness holds in

$$0 \to \operatorname{Hom}(C, G) \xrightarrow{k^{\#}} \operatorname{Hom}(B, G) \xrightarrow{j^{\#}} \operatorname{Hom}(A, G) \to \operatorname{Ext}(C, G)$$

$$\xrightarrow{h^{*}} \operatorname{Ext}(B, G) \xrightarrow{j^{*}} \operatorname{Ext}(A, G) \to 0.$$

Since we have defined $S^n(X, A; G) = \operatorname{Hom}(S_n(X, A), G)$, it is useful to adopt the following notation: if ϕ is in $S^n(X, A; G)$ and c is in $S_n(X, A)$, then the value of ϕ on c is the element of G denoted by $\langle \phi, c \rangle$. Note that this pairing is bilinear in the sense that

$$\langle \phi_1 + \phi_2, c \rangle = \langle \phi_1, c \rangle + \langle \phi_2, c \rangle$$

and

$$\langle \phi, c_1 + c_2 \rangle = \langle \phi, c_1 \rangle + \langle \phi, c_2 \rangle.$$

In particular, for any integer n we have $\langle \phi, nc \rangle = \langle n\phi, c \rangle$. Thus the pairing produces a homomorphism

$$S^n(X, A; G) \otimes S_n(X, A) \to G$$

for each n. This homomorphism will be studied in more detail in the next chapter.

In this notation the boundary and coboundary operators are adjoint; that is,

$$\langle \phi, \partial c \rangle = \langle \delta\phi, c \rangle.$$

(In a somewhat different setting this is called the fundamental theorem of calculus.) Furthermore, if $f: (X, A) \to (Y, B)$ is a map of pairs, $\phi \in S^n(Y, B; G)$ and $c \in S_n(X, A)$, then

$$\langle \phi, f_{\#}(c) \rangle = \langle f^{\#}(\phi), c \rangle.$$

The cochain $\phi \in S^n(X, A; G)$ is a cocycle if and only if

$$\langle \delta\phi, c' \rangle = 0 \qquad \text{for all} \quad c' \in S_{n+1}(X, A),$$

or equivalently, if

$$\langle \phi, \partial c' \rangle = 0.$$

Thus, ϕ is a cocycle if and only if ϕ annihilates $B_n(X, A)$.

On the other hand, suppose that $\phi = \delta\phi'$ is a coboundary, where $\phi' \in S^{n-1}(X, A; G)$. Then

$$\langle \phi, c \rangle = \langle \delta\phi', c \rangle = \langle \phi', \partial c \rangle$$

so that if ϕ is a coboundary, then ϕ annihilates $Z_n(X, A)$.

Now let $x \in H^n(X, A; G)$ be represented by a cocycle ϕ and $y \in H_n(X, A)$ be represented by a cycle c. Then we define a pairing

$$\langle \ , \ \rangle: \quad H^n(X, A; G) \otimes H_n(X, A) \to G$$

by $\langle x, y \rangle = \langle \phi, c \rangle$. To see that this is well defined, let $\phi + \delta\phi'$ and $c + \partial c'$ be other choices for representatives of x and y. Then

$$\langle \phi + \delta\phi', c + \partial c' \rangle = \langle \phi, c \rangle + \langle \delta\phi', c + \partial c' \rangle + \langle \phi, \partial c' \rangle$$
$$= \langle \phi, c \rangle + \langle \phi', \partial c + \partial\partial c' \rangle + \langle \delta\phi, c' \rangle$$
$$= \langle \phi, c \rangle.$$

This pairing is called the *Kronecker index* and may be viewed as a homomorphism

$$\alpha: \quad H^n(X, A; G) \to \operatorname{Hom}(H_n(X, A), G).$$

For example, consider the zero-dimensional cohomology of a space X. $S^0(X; G) = \operatorname{Hom}(S_0(X), G)$ and $S_0(X)$ may be identified with the free abelian group generated by the points of X. Since any homomorphism defined on $S_0(X)$ is determined by its value on the basis, we may identify $S^0(X; G)$ with the set of all functions from X to G.

Of course, $B^0(X; G) = 0$, so the cohomology may be determined by identifying the group of 0-cocycles. Note that ϕ will be a 0-cocycle if and only if $\langle \delta\phi, c \rangle = \langle \phi, \partial c \rangle = 0$ for all c in $S_1(X)$. This will be true if and only if $\phi(\sigma(1)) = \phi(\sigma(0))$ for every path σ in X. Therefore, we have identified $H^0(X; G) = Z^0(X; G)$ with the set of all functions from X to G which are constant on the path components of X. From this description we have the following.

3.11 PROPOSITION If X is a topological space and $\{X_\alpha\}_{\alpha \in \Lambda}$ is the decomposition of X into its path components, then

$$H^0(X; G) \approx \prod_{\alpha \in \Lambda} G_\alpha,$$

the direct product of copies of G, one for each path component of X. $\square$

There is a natural embedding of G in $H^0(X; G)$ defined by sending $g \in G$ into the function from X to G having constant value g. If p is a point and $\pi: X \to p$ is the map of X to p, then

$$\pi^*: \quad H^0(p; G) \to H^0(X; G)$$

maps $H^0(p; G) \approx G$ isomorphically onto this embedded copy of G. As for the case of homology we define

$$\tilde{H}^*(X; G) = H^*(X; G)/\text{im } \pi^*,$$

the *reduced* cohomology group of X with coefficients in G.

Essentially all of the results we have established previously for homology carry over in dual form to cohomology. For example, we have the following.

3.12   THEOREM   Let (X, A) be a pair of spaces and $U \subseteq A$ a subset with $\bar{U} \subseteq \text{Int } A$. Then the inclusion map of pairs

$$i: \quad (X - U, A - U) \to (X, A)$$

induces an isomorphism

$$i^*: \quad H^*(X, A; G) \to H^*(X - U, A - U; G).$$

PROOF   We have observed that

$$i_{\#}: \quad S_*(X - U, A - U) \to S_*(X, A)$$

induces an isomorphism on homology groups (Theorem 2.11). The following exercise implies that $i_{\#}$ is a chain homotopy equivalence. Thus,

$$i^{\#}: \quad S^*(X, A; G) \to S^*(X - U, A - U; G)$$

is a cochain homotopy equivalence and i^* is an isomorphism. $\square$

Exercise 4   If $f: C \to D$ is a chain map of degree zero between free chain complexes, such that f induces an isomorphism of homology groups, then f is a chain homotopy equivalence. [Hint: Take the algebraic mapping cone C_f of f (see page 152). Use the fact that C_f has trivial homology to construct the homotopies.]

3.13   THEOREM   If $f, g: (X, A) \to (Y, B)$ are homotopic as maps of pairs, then the induced homomorphisms

$$f^*, g^*: \quad H^*(Y, B; G) \to H^*(X, A; G)$$

are equal.

PROOF   We showed in Theorem 2.10 that

$$f_\#, g_\#: \quad S_*(X, A) \to S_*(Y, B)$$

are chain homotopic. Therefore

$$f^\#, g^\#: \quad S^*(Y, B; G) \to S^*(X, A; G)$$

are cochain homotopic, and it follows that $f^* = g^*$. $\square$

Exercise 5   Formulate and prove the Mayer–Vietoris sequence for singular cohomology.

Exercise 6   State the Eilenberg–Steenrod axioms for cohomology and prove the uniqueness theorem in the category of finite CW complexes.

Example   We want to compute $H^*(S^n, x_0; G)$. First note that $H^k(S^0, x_0; G)$ is isomorphic by excision to

$$H^k(\text{pt}; G) \approx \begin{cases} G & \text{for} \quad k = 0 \\ 0 & \text{otherwise.} \end{cases}$$

As usual (Figure 3.1) we decompose the n-sphere into its upper cap E_+^n

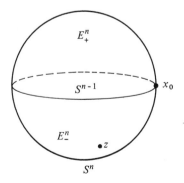

Figure 3.1

and lower cap E_-^n with $x_0 \in S^{n-1} = E_+^n \cap E_-^n$. Let z be a point in the interior of E_-^n.

Note that since the inclusions $x_0 \to E_+^n$ and $x_0 \to E_-^n$ are homotopy equivalences, the relative cohomology groups $H^*(E_+^n, x_0; G)$ and $H^*(E_-^n, x_0; G)$ are both zero. Thus, in the exact cohomology sequence of the triple (S^n, E_-^n, x_0) we have an isomorphism

$$H^k(S^n, E_-^n; G) \xrightarrow{\approx} H^k(S^n, x_0; G).$$

The point z may be excised from the pair (S^n, E_-^n) to give an isomorphism

$$H^k(S^n, E_-^n; G) \xrightarrow{\approx} H^k(S^n - z, E_-^n - z; G).$$

Now the pair $(S^n - z, E_-^n - z)$ may be mapped by a relative homeomorphism to the pair (E_+^n, S^{n-1}) so that we have an isomorphism

$$H^k(S^n - z, E_-^n - z; G) \approx H^k(E_+^n, S^{n-1}; G).$$

Finally, the exact sequence of the triple (E_+^n, S^{n-1}, x_0) yields an isomorphism

$$H^{k-1}(S^{n-1}, x_0; G) \xrightarrow{\approx} H^k(E_+^n, S^{n-1}; G).$$

Thus

$$H^k(S^n, x_0; G) \approx H^{k-1}(S^{n-1}, x_0; G),$$

which completes the inductive step to prove that

$$H^k(S^n, x_0; G) \approx \begin{cases} G & \text{for } k = n \\ 0 & \text{otherwise.} \end{cases}$$

All of these similarities between homology and cohomology might lead one to ask: why bother? There may be many answers to this question; we briefly cite only three:

(i) When the coefficient group is also a ring, the cohomology of a space may be given a natural ring structure. (This is not true for homology groups.) This additional algebraic structure gives us another topological invariant.

(ii) Cohomology theory is the natural setting for "characteristic classes." These are particular cohomology classes, arising in the study of fiber bundles, which have many applications, particularly to the topology of manifolds.

(iii)   There are "cohomology operations," naturally occurring transformations in cohomology theory, which have many applications in homotopy
theory.

In Chapter 4 we will define the ring structure in (i) while studying the
relationships between homology and cohomology theory. The topics in (ii)
and (iii) are more advanced and will not be dealt with here. Perhaps the
best source for topic (ii) is Milnor [1957]. The original source for topic (iii)
is Steenrod and Epstein [1962]; a more recent book is by Mosher and
Tangora [1968].

We close this chapter with an extension of the universal coefficient theorem
which establishes the first basic connection between homology and cohomology groups.

3.14   THEOREM   Given a pair (X, A) of spaces and an abelian group G,
there exists a split exact sequence

$$0 \to \mathrm{Ext}(H_{n-1}(X, A), G) \to H^n(X, A; G) \xrightarrow{\alpha} \mathrm{Hom}(H_n(X, A), G) \to 0.$$

Exercise 7   Prove Theorem 3.14. $\square$

chapter 4

PRODUCTS

In this chapter we introduce the theory of products in homology and cohomology. Following the Künneth formula for free chain complexes, we state and prove the acyclic model theorem. This is applied to establish the Eilenberg–Zilber theorem and the resulting external products in homology and cohomology. When the coefficient group is a ring R, it is shown that the cohomology external product may be refined to the cup product, giving the cohomology group the structure of an R-algebra. This structure is computed for the torus by introducing the Alexander–Whitney diagonal approximation. Also, a cup product definition of the Hopf invariant is given. Finally, the cap product between homology and cohomology is defined in anticipation of Chapter 5.

Suppose that $C = \{C_n, \partial\}$ and $D = \{D_n, \partial\}$ are chain complexes. In Chapter 3 we discussed the formation of a new chain complex by tensoring a given chain complex with an abelian group. We now want to generalize this to give a procedure for tensoring two chain complexes to form a new chain complex.

Define a chain complex $C \otimes D$ by setting

$$(C \otimes D)_n = \sum_k C_k \otimes D_{n-k}.$$

The boundary operator on a direct summand

$$\partial: \quad C_p \otimes D_q \rightarrow C_{p-1} \otimes D_q \oplus C_p \otimes D_{q-1}$$

is given by the formula

$$\partial(c \otimes d) = \partial c \otimes d + (-1)^p c \otimes \partial d.$$

To check that this gives a chain complex, note that

$$\begin{aligned}
\partial(\partial(c \otimes d)) &= \partial(\partial c \otimes d + (-1)^p c \otimes \partial d) \\
&= \partial\partial c \otimes d + (-1)^{p-1} \partial c \otimes \partial d + (-1)^p \partial c \otimes \partial d \\
&\quad + (-1)^{2p}(c \otimes \partial\partial d) \\
&= 0.
\end{aligned}$$

Since the elements $(c \otimes d)$ generate $C \otimes D$, it follows that $\partial \circ \partial = 0$.

Note that if $f: C \rightarrow C'$ and $g: D \rightarrow D'$ are chain maps between chain complexes, there is an associated chain map

$$f \otimes g: \quad C \otimes D \rightarrow C' \otimes D'$$

characterized by $f \otimes g(c \otimes d) = f(c) \otimes g(d)$.

Now suppose that C is a free chain complex. The exact sequence

$$0 \rightarrow Z_n(C) \xrightarrow{\alpha} C_n \xrightarrow{\partial} B_{n-1}(C) \rightarrow 0,$$

where α is the inclusion, must split because $B_{n-1}(C)$ is free. Thus, there exists a homomorphism

$$\phi: \quad C_n \rightarrow Z_n(C)$$

which is just projection onto a direct summand, that is, $\phi \circ \alpha = $ identity on $Z_n(C)$.

We consider the graded groups $Z_*(C)$, $B_*(C)$, and $H_*(C)$ to be chain complexes in which the boundary operators are all identically zero. Denote by Φ the composition $\Phi = \pi \circ \phi$,

$$C_n \xrightarrow{\phi} Z_n(C) \xrightarrow{\pi} H_n(C),$$

where π is the quotient map. Then Φ is a chain map between chain complexes because

$$\Phi(\partial c) = \pi(\partial c) = 0 = \partial\Phi(c).$$

4.1 THEOREM If C and D are free chain complexes, the chain map

$$\Phi \otimes \text{id}: \quad C \otimes D \to H_*(C) \otimes D$$

induces an isomorphism

$$(\Phi \otimes d)_*: \quad H_n(C \otimes D) \to H_n(H_*(C) \otimes D).$$

PROOF Recall the exact sequence of chain complexes and chain maps

$$0 \to Z_*(C) \xrightarrow{\alpha} C \xrightarrow{\partial} B_*(C) \to 0,$$

where ∂ has degree -1. Since the sequence splits, we may tensor with the chain complex D and preserve exactness. This yields an exact sequence of chain complexes and chain maps

$$0 \to Z_*(C) \otimes D \xrightarrow{\alpha \otimes \text{id}} C \otimes D \xrightarrow{\partial \otimes \text{id}} B_*(C) \otimes D \to 0$$

and thus an exact homology sequence

$$\cdots \to H_n(Z_*(C) \otimes D) \xrightarrow{(\alpha \otimes \text{id})_*} H_n(C \otimes D) \xrightarrow{(\partial \otimes \text{id})_*} H_{n-1}(B_*(C) \otimes D)$$
$$\xrightarrow{\Delta} H_{n-1}(Z_*(C) \otimes D) \to \cdots,$$

where $(\partial \otimes \text{id})_*$ has degree -1 and Δ is the connecting homomorphism. On the other hand, the short exact sequence

$$0 \to B_*(C) \xrightarrow{\beta} Z_*(C) \xrightarrow{\pi} H_*(C) \to 0$$

of chain complexes need not split. However, since D is free, exactness will be preserved in

$$0 \to B_*(C) \otimes D \xrightarrow{\beta \otimes \text{id}} Z_*(C) \otimes D \xrightarrow{\pi \otimes \text{id}} H_*(C) \otimes D \to 0.$$

Passing to the homology groups of these complexes we have the long exact sequence

$$\cdots \to H_n(B_*(C) \otimes D) \xrightarrow{(\beta \otimes \text{id})_*} H_n(Z_*(C) \otimes D) \xrightarrow{(\pi \otimes \text{id})_*} H_n(H_*(C) \otimes D)$$
$$\xrightarrow{\Delta'} H_{n-1}(B_*(C) \otimes D) \to \cdots.$$

These two long exact sequences

$$H_n(B_*(C) \otimes D) \xrightarrow{\;\Delta\;} H_n(Z_*(C) \otimes D) \xrightarrow{(\alpha \otimes \mathrm{id})_*} H_n(C \otimes D)$$

$$\downarrow = \qquad\qquad \downarrow = \qquad\qquad \downarrow (\Phi \otimes \mathrm{id})_*$$

$$H_n(B_*(C) \otimes D) \xrightarrow{(\beta \otimes \mathrm{id})_*} H_n(Z_*(C) \otimes D) \xrightarrow{(\pi \otimes \mathrm{id})_*} H_n(H_*(C) \otimes D)$$

$$\xrightarrow{(\partial \otimes \mathrm{id})_*} H_{n-1}(B_*(C) \otimes D) \xrightarrow{\;\Delta\;} H_{n-1}(Z_*(C) \otimes D)$$

$$\downarrow = \qquad\qquad\qquad \downarrow =$$

$$\xrightarrow{\;\Delta'\;} H_{n-1}(B_*(C) \otimes D) \xrightarrow{(\beta \otimes \mathrm{id})_*} H_{n-1}(Z_*(C) \otimes D)$$

may be related in such a way that each rectangle commutes up to sign (see the following exercise). Now the proof of the five lemma (Exercise 4, Chapter 2) only required commutativity up to sign; hence, we apply the five lemma to conclude that

$$(\Phi \otimes \mathrm{id})_*: \quad H_n(C \otimes D) \xrightarrow{\;\approx\;} H_n(H_*(C) \otimes D)$$

is an isomorphism. This completes the proof. $\square$

Exercise 1 Show that in the diagram in the preceding proof each rectangle commutes up to sign.

This proposition reduces the problem of computing the homology of the chain complex $C \otimes D$ to computing the homology of the simpler complex $H_*(C) \otimes D$. Note that if $c \otimes d \in H_p(C) \otimes D_q$, then $\partial(c \otimes d) = (-1)^p c \otimes \partial d$, so that up to sign the boundary operator is just

$$\mathrm{id} \otimes \partial: \quad H_p(C) \otimes D_q \to H_p(C) \otimes D_{q-1}.$$

Therefore, for p fixed $H_p(C) \otimes D$ is a subcomplex of $H_*(C) \otimes D$, in fact a direct summand, and we conclude that

$$H_n(H_*(C) \otimes D) = \sum_p H_n(H_p(C) \otimes D).$$

Now if two boundary operators differ by sign only, it is evident that they produce the same homology groups. Thus, we may assume that the boundary operator in the chain complex $H_p(C) \otimes D$ is $\mathrm{id} \otimes \partial$. Note that the n-dimensional component of this complex is $H_p(C) \otimes D_{n-p}$.

Since D is a free chain complex, we are in a position to apply the universal coefficient theorem, Theorem 3.6, to the chain complex $H_p(C) \otimes D$. Thus

$$H_n(H_p(C) \otimes D) \approx H_p(C) \otimes H_{n-p}(D) \oplus \text{Tor}(H_p(C), H_{n-p-1}(D)).$$

Summing these over all values of p, we have completed the proof of the *Künneth formula* for free chain complexes:

4.2 COROLLARY If C and D are free chain complexes, then

$$H_n(C \otimes D) \approx \sum_{p+q=n} H_p(C) \otimes H_q(D) \oplus \sum_{r+s=n-1} \text{Tor}(H_r(C), H_s(D)). \qquad \square$$

Example Suppose $c \in Z_p(C)$ but c is not a boundary. Suppose further that $r \cdot c = \partial c'$ for some $c' \in C_{p+1}$ and some minimal integer $r > 0$, so that c represents a homology class of order r. Similarly let $d \in Z_q(D)$ represent a homology class of order r so that $rd = \partial d'$ for some $d' \in D_{q+1}$. Then in $(C \otimes D)_{p+q+1}$ the element $(c' \otimes d - (-1)^p c \otimes d')$ is a cycle because

$$\partial(c' \otimes d - (-1)^p c \otimes d') = \partial c' \otimes d + (-1)^{p+1} c' \otimes \partial d - (-1)^p \partial c \otimes d'$$
$$- (-1)^{2p} c \otimes \partial d'$$
$$= rc \otimes d - c \otimes rd$$
$$= r(c \otimes d - c \otimes d)$$
$$= 0.$$

In this way torsion common to $H_p(C)$ and $H_q(D)$ produces homology classes in $H_{p+q+1}(C \otimes D)$.

Given spaces X and Y, the Künneth formula of Corollary 4.2 may be applied to the singular chain complexes $S_*(X)$ and $S_*(Y)$ to give the isomorphism

$$H_n(S_*(X) \otimes S_*(Y)) \approx \sum_{p+q=n} H_p(X) \otimes H_q(Y) \oplus \sum_{r+s=n-1} \text{Tor}(H_r(X), H_s(Y)).$$

We turn now to the problem of relating $H_n(S_*(X) \otimes S_*(Y))$ to $H_n(X \times Y)$, the homology of the cartesian product of X and Y.

The solution of this problem will be stated in terms of the acyclic model theorem, a useful tool in homological algebra. To put this result in its proper setting we require a number of definitions. A *category* $\mathcal{C}$ is

(a) a class of *objects*,

(b) for every ordered pair of objects a set hom(X, Y), of *morphisms* viewed as functions with domain X and range Y,

such that whenever $f: X \to Y$ and $g: Y \to Z$ are morphisms, there is an element $g \circ f$ in hom(X, Z). These are required to satisfy the following axioms:

1. *Associativity*: $(h \circ g) \circ f = h \circ (g \circ f)$.
2. *Identity*: For every object Y there is an element $1_Y \in$ hom(Y, Y) such that if $f: X \to Y$ and $g: Y \to Z$ are morphisms, then $1_Y \circ f = f$ and $g \circ 1_Y = g$.

Examples (i) The category whose objects are sets and whose morphisms are functions.

(ii) The category of abelian groups and homomorphisms.

(iii) The category of topological spaces and continuous functions.

(iv) The category of pairs of spaces and maps of pairs.

(v) The category of chain complexes and chain maps.

If $\mathcal{C}$ and $\mathcal{D}$ are categories, a *covariant functor* $T: \mathcal{C} \to \mathcal{D}$ is a function that assigns to each object X in $\mathcal{C}$ an object $T(X)$ in $\mathcal{D}$ and to each morphism $f: X \to Y$ a morphism $T(f): T(X) \to T(Y)$ such that

(a) $T(1_Y) = 1_{T(Y)}$,
(b) $T(f \circ g) = T(f) \circ T(g)$.

A functor K is *contravariant* if for $f: X \to Y$, $K(f): K(Y) \to K(X)$ and

(a') $K(1_Y) = 1_{K(Y)}$,
(b') $K(f \circ g) = K(g) \circ K(f)$.

Examples (1) The correspondence $(X, A) \to S_*(X, A)$ and $[f: (X, A) \to (Y, B)] \to [f_\#: S_*(X, A) \to S_*(Y, B)]$ is a covariant functor from Category (iv) above to Category (v).

(2) The correspondence $X \to H^n(X; G)$ and $[f: X \to Y] \to [f^*: H^n(Y; G) \to H^n(X; G)]$ is a contravariant functor from Category (iii) to Category (ii).

Suppose that $\mathcal{C}$ and $\mathcal{D}$ are categories and $T_1, T_2: \mathcal{C} \to \mathcal{D}$ are covariant functors. A *natural transformation* $\tau: T_1 \to T_2$ is a function which assigns to each object X in $\mathcal{C}$ a morphism $\tau(X): T_1(X) \to T_2(X)$ in $\mathcal{D}$ such that commutativity holds in

$$T_1(X) \xrightarrow{T_1(f)} T_1(Y)$$

$$\downarrow{\tau(X)} \qquad \downarrow{\tau(Y)}$$

$$T_2(X) \xrightarrow{T_2(f)} T_2(Y)$$

whenever $f: X \to Y$ is a morphism in $\mathcal{C}$.

Now fix a category $\mathcal{C}$. Suppose that $\mathfrak{M} = \{M_\alpha\}_{\alpha \in \Lambda}$ is a specified collection of objects in $\mathcal{C}$. $\mathfrak{M}$ will be called the *models* of $\mathcal{C}$. A functor T from $\mathcal{C}$ to the category of abelian groups and homomorphisms is *free with respect to the models* $\mathfrak{M}$ if there exists an element $e_\alpha \in T(M_\alpha)$ for each α such that for every X in $\mathcal{C}$ the set

$$\{T(f)(e_\alpha) \mid \alpha \in \Lambda, f \in \text{hom}(M_\alpha, X)\}$$

is a basis for $T(X)$ as a free abelian group. A functor T from $\mathcal{C}$ to the category of graded abelian groups is free with respect to the models $\mathfrak{M}$ if each T_n is, where T_n is the nth component of T.

4.3   THEOREM (*Acyclic model theorem*)   Let $\mathcal{C}$ be a category with models $\mathfrak{M}$ and T, T' covariant functors from $\mathcal{C}$ to the category of chain complexes and chain maps, such that $T_n = 0 = T_n'$ for $n < 0$ and T is free with models $\mathfrak{M}$. Suppose further that $H_i(T'(M_\alpha)) = 0$ for $i > 0$ and $M_\alpha \in \mathfrak{M}$. If there is a natural transformation

$$\Phi: \quad H_0(T) \to H_0(T'),$$

then there is a natural transformation

$$\phi: \quad T \to T'$$

which induces Φ, and furthermore any two such ϕ are naturally chain homotopic.

PROOF   By the hypothesis $T_0(M_\alpha)$ and $T_0'(M_\alpha)$ are the respective cycle groups in dimension zero. Thus, there are epimorphisms π and π' onto the homology groups:

$$T_0(M_\alpha) \xrightarrow{\pi} H_0(T(M_\alpha))$$

$$\downarrow{\phi} \qquad \qquad \downarrow{\Phi}$$

$$T_0'(M_\alpha) \xrightarrow{\pi'} H_0(T'(M_\alpha))$$

Since T_0 is free with models $\mathfrak{M}$, there is for each α a prescribed element $e_\alpha^0 \in T_0(M_\alpha)$. So for each α we choose an element $\phi(e_\alpha^0) \in T_0'(M_\alpha)$ such that $\pi' \circ \phi(e_\alpha^0) = \Phi \circ \pi(e_\alpha^0)$.

Let $f\colon M_\alpha \to X$ be a morphism in $\mathcal{C}$. Then $T(f)(e_\alpha^0)$ is a basis element in $T_0(X)$ and we define $\phi(T(f)(e_\alpha^0)) = T'(f)(\phi(e_\alpha^0))$. This defines ϕ on the basis elements of the free abelian group $T_0(X)$ so there is a unique extension to a homomorphism

$$\phi\colon \quad T_0(X) \to T_0'(X).$$

To check that ϕ induces the original Φ on zero dimensional homology, we must show that the front face of the following diagram commutes:

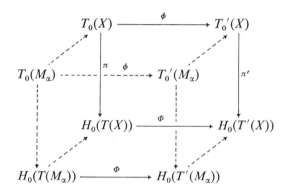

The bottom face commutes by the naturality of Φ. The left and right faces commute since $T(f)$ and $T'(f)$ are chain maps. The back face and the top face commute by definition; thus, the front face must also commute.

Since T_1 is free with models $\mathfrak{M}$, there is for each α a prescribed element $e_\alpha^1 \in T_1(M_\alpha)$. From the above, $\phi(\partial e_\alpha^1)$ is a well-defined element of $T_0'(M_\alpha)$. Moreover, since ϕ induces Φ on zero-dimensional homology, $\phi(\partial e_\alpha^1)$ must be a boundary in $T_0'(M_\alpha)$. So let $c \in T_1'(M_\alpha)$ with $\partial c = \phi(\partial e_\alpha^1)$ and define $\phi(e_\alpha^1) = c$. Using the above technique we extend ϕ to a homomorphism $\phi\colon T_1(X) \to T_1'(X)$ for each object X in $\mathcal{C}$.

Suppose ϕ is defined in dimensions less than n, and consider the set $\{e_\alpha^n \mid e_\alpha^n \in T_n(M_\alpha)\}$ given by the fact that T_n is free with models $\mathfrak{M}$. By the inductive hypothesis $\phi(\partial e_\alpha^n)$ is a well-defined element of $T_{n-1}'(M_\alpha)$. Since it is a cycle and T' is acyclic in positive dimensions, it is also a boundary. So define $\phi(e_\alpha^n)$ to be an element of $T_n'(M_\alpha)$ whose boundary is $\phi(\partial e_\alpha^n)$. Once again ϕ may be extended using the fact that T is free with models $\mathfrak{M}$. This defines ϕ on $T(X)$ for all objects X in $\mathcal{C}$.

Exercise 2   Show that for each X in $\mathcal{C}$, $\phi \colon T(X) \to T'(X)$ is a chain map, and for each morphism $f \colon X \to Y$, $\phi \circ T(f) = T'(f) \circ \phi$.

This defines the natural transformation $\phi \colon T \to T'$. Suppose now that $\phi' \colon T \to T'$ is another such natural transformation, inducing Φ on zero-dimensional homology. For each object X in $\mathcal{C}$ we must construct a chain homotopy $\mathfrak{I} \colon T(X) \to T'(X)$, which is natural with respect to morphisms in $\mathcal{C}$, having

$$\partial \mathfrak{I} + \mathfrak{I} \partial = \phi - \phi'.$$

We define $\mathfrak{I}$ inductively. Suppose that it has been defined in dimensions less than n, and recall that $T_n(X)$ has basis $\{T(f)(e_\alpha{}^n)\}$ as a free abelian group. For $n > 0$ the element

$$\phi(e_\alpha{}^n) - \phi'(e_\alpha{}^n) - \mathfrak{I}(\partial e_\alpha{}^n)$$

is a cycle because

$$\partial \phi(e_\alpha{}^n) - \partial \phi'(e_\alpha{}^n) - \partial \mathfrak{I}(\partial e_\alpha{}^n)$$
$$= \phi \partial e_\alpha{}^n - \phi' \partial e_\alpha{}^n - (-\mathfrak{I} \partial \partial e_\alpha{}^n + \phi \partial e_\alpha{}^n - \phi' \partial e_\alpha{}^n)$$
$$= 0.$$

Since T' is acyclic in positive dimensions, this cycle must bound, so define $\mathfrak{I}(e_\alpha{}^n)$ to be an element of $T'_{n+1}(M_\alpha)$ whose boundary is $(\phi(e_\alpha{}^n) - \phi'(e_\alpha{}^n) - \mathfrak{I}(\partial e_\alpha{}^n))$. Again we extend $\mathfrak{I}$ to be defined on $T(X)$ for all objects X in $\mathcal{C}$ by using the fact that T is free with models $\mathfrak{M}$. This same technique will work for the case $n = 0$ because the cycle $\phi(e_\alpha{}^0) - \phi'(e_\alpha{}^0)$ must bound. This is a consequence of the fact that ϕ and ϕ' induce the same homomorphism (Φ) on zero-dimensional homology.

This $\mathfrak{I}$ gives the desired chain homotopy and is natural with respect to morphisms of $\mathcal{C}$, so the proof is complete. $\square$

NOTE   The technique in the proof of Theorem 4.3 is essentially the same as that used in Theorem 1.10 and Appendix I, although the former is in a more general context.

Exercise 3   Re-prove Theorem 1.10 as a corollary to the acyclic model theorem.

We now want to apply this theorem to relate the homology of the chain complex $S_*(X \times Y)$ to the homology of $S_*(X) \otimes S_*(Y)$. Let $\mathcal{C}$ be the category of topological spaces and continuous functions. (This may easily be generalized to the category of pairs of spaces and maps of pairs.) Denote by $\mathcal{C} \times \mathcal{C}$ the category whose objects are ordered pairs (X, Y) of objects in $\mathcal{C}$ and whose morphisms are ordered pairs (f, f') of morphisms in $\mathcal{C}$ with, $f: X \to X'$ and $f': Y \to Y'$. Let $\mathfrak{M}$ be the set of all pairs (σ^p, σ^q), $p, q \geq 0$ in $\mathcal{C} \times \mathcal{C}$ where σ^k is the standard k-simplex. Define two functors from $\mathcal{C} \times \mathcal{C}$ to the category of chain complexes and chain maps by

$$T(X, Y) = S_*(X \times Y) \quad \text{and} \quad T'(X, Y) = S_*(X) \otimes S_*(Y).$$

Both of these functors are free with models $\mathfrak{M}$. Furthermore, both have models acyclic in positive dimensions.

The path components of $X \times Y$ are of the form $C \times D$, where C and D are path components of X and Y, respectively. As a result there is a natural isomorphism

$$H_0(X \times Y) \xrightarrow{\Phi} H_0(S_*(X) \otimes S_*(Y))$$

because $H_0(S_*(X) \otimes S_*(Y)) \approx H_0(X) \otimes H_0(Y)$ by the Kunneth formula of Corollary 4.2.

From the natural transformations Φ and Φ^{-1} we apply the acyclic model theorem, Theorem 4.3, in each direction to conclude that there exist chain maps

$$\phi: \quad S_*(X \times Y) \to S_*(X) \otimes S_*(Y)$$

and

$$\bar{\phi}: \quad S_*(X) \otimes S_*(Y) \to S_*(X \times Y)$$

that induce Φ and Φ^{-1}, respectively, in dimension zero.

Thus, $\phi \circ \bar{\phi}$ is a chain map from $S_*(X) \otimes S_*(Y)$ to itself inducing the identity on zero-dimensional homology. But the identity chain map also has this property, so by Theorem 4.3, $\phi \circ \bar{\phi}$ is chain homotopic to the identity. Similarly the composition $\bar{\phi} \circ \phi$ is chain homotopic to the identity on $S_*(X \times Y)$. Therefore

$$\phi_*: \quad H_*(X \times Y) \to H_*(S_*(X) \otimes S_*(Y))$$

is an isomorphism with inverse $\bar{\phi}_*$. This completes the proof of the *Eilenberg–Zilber theorem*:

4.4 THEOREM   For any spaces X and Y and any integer k there is an isomorphism

$$\phi_*: \quad H_k(X \times Y) \to H_k(S_*(X) \otimes S_*(Y)). \quad \square$$

By combining Theorem 4.4 and Corollary 4.2 we have established the Künneth formula for singular homology theory:

4.5 THEOREM If X and Y are spaces, there is a natural isomorphism

$$H_n(X \times Y) \approx \sum_{p+q=n} H_p(X) \otimes H_q(Y) \oplus \sum_{r+s=n-1} \mathrm{Tor}(H_r(X), H_s(Y))$$

for each n. $\square$

Suppose now that we have fixed a natural chain map

$$\phi: \quad S_*(X \times Y) \to S_*(X) \otimes S_*(Y)$$

for any spaces X and Y with the above properties. The composition

$$H_p(X) \otimes H_q(Y) \to H_{p+q}(S_*(X) \otimes S_*(Y)) \xrightarrow{\phi_*^{-1}} H_{p+q}(X \times Y),$$

where the first homomorphism takes $\{x\} \otimes \{y\}$ into $\{x \otimes y\}$, is called the *homology external product*. The image of $\{x\} \otimes \{y\}$ under the composition is usually denoted $\{x\} \times \{y\}$. From the Künneth formula we may conclude that this is a monomorphism for any choice of p and q. In fact the Künneth formula for singular homology may be restated as a split exact sequence

$$0 \to \sum_{p+q=n} H_p(X) \otimes H_q(Y) \to H_n(X \times Y) \to \sum_{r+s=n-1} \mathrm{Tor}(H_r(X), H_s(Y)) \to 0,$$

where the monomorphism is given by the external product.

Our primary purpose now is to construct the analog of this in cohomology, that is, a product

$$H^p(X; G_1) \otimes H^q(Y; G_2) \dashrightarrow H^{p+q}(X \times Y; G_1 \otimes G_2).$$

If $\alpha \in S^p(X; G_1)$ and $\beta \in S^q(Y; G_2)$, then $\alpha: S_p(X) \to G_1$ and $\beta: S_q(Y) \to G_2$ are homomorphisms. Denote by $\alpha \times \beta$ the homomorphism given by the composition

$$S_{p+q}(X \times Y) \xrightarrow{\phi} S_*(X) \otimes S_*(Y) \xrightarrow{\alpha \otimes \beta} G_1 \otimes G_2,$$

where $\alpha \otimes \beta$ is defined to be zero on any term not lying in $S_p(X) \otimes S_q(Y)$. Thus, $\alpha \times \beta \in S^{p+q}(X \times Y; G_1 \otimes G_2)$. This defines an external product on cochains

$$S^p(X; G_1) \otimes S^q(Y; G_2) \to S^{p+q}(X \times Y; G_1 \otimes G_2).$$

4.6 PROPOSITION If $\alpha \in S^p(X; G_1)$ and $\beta \in S^q(Y; G_2)$ are cochains and $\alpha \times \beta \in S^{p+q}(X \times Y; G_1 \otimes G_2)$ is their external product, then

$$\delta(\alpha \times \beta) = (\delta\alpha) \times \beta + (-1)^p \alpha \times \delta\beta.$$

(This is the *derivation formula* for cochains.)

PROOF The diagram

$$
\begin{array}{ccc}
S_{p+q+1}(X \times Y) & \xrightarrow{\ \phi\ } & S_*(X) \otimes S_*(Y) \\
\downarrow{\scriptstyle\partial} & & \downarrow{\scriptstyle\partial} \\
S_{p+q}(X \times Y) & \xrightarrow{\ \phi\ } & S_*(X) \otimes S_*(Y) \xrightarrow{\ \alpha \otimes \beta\ } G_1 \otimes G_2
\end{array}
$$

commutes since ϕ is a chain map. Thus

$$\delta(\alpha \times \beta) = (\alpha \otimes \beta \circ \phi) \circ \partial = (\alpha \otimes \beta) \circ \partial \circ \phi.$$

On the other hand

$$(\delta\alpha) \times \beta = ((\delta\alpha) \otimes \beta) \circ \phi \qquad \text{and} \qquad \alpha \times \delta\beta = (\alpha \otimes \delta\beta) \circ \phi.$$

Therefore, it is sufficient to check the behavior of these three homomorphisms on the image of ϕ.

Let $e \otimes c$ be a basis element of $S_*(X) \otimes S_*(Y)$. Then $(\alpha \otimes \beta) \circ \partial$ will be zero on $e \otimes c$ unless

(i) $e \in S_{p+1}(X)$ and $c \in S_q(Y)$ or
(ii) $e \in S_p(X)$ and $c \in S_{q+1}(Y)$.

In the first case

$$
\begin{aligned}
(\alpha \otimes \beta) \circ \partial(e \otimes c) &= (\alpha \otimes \beta)(\partial e \otimes c + (-1)^{p+1} e \otimes \partial c) \\
&= \alpha(\partial e) \otimes \beta(c) + 0 \\
&= ((\delta\alpha) \otimes \beta)(e \otimes c).
\end{aligned}
$$

In the second case

$$
\begin{aligned}
(\alpha \otimes \beta) \circ \partial(e \otimes c) &= (\alpha \otimes \beta)(\partial e \otimes c + (-1)^p e \otimes \partial c) \\
&= (-1)^p \alpha(e) \otimes \beta(\partial c) \\
&= (-1)^p (\alpha \otimes \delta\beta)(e \otimes c).
\end{aligned}
$$

Since all three homomorphisms will be zero on any basis element not of the form (i) or (ii), we conclude that $\delta(\alpha \times \beta) = (\delta\alpha) \times \beta + (-1)^p(\alpha \times \delta\beta)$. $\square$

4.7 COROLLARY This induces a well-defined external product on cohomology groups

$$H^p(X; G_1) \otimes H^q(Y; G_2) \to H^{p+q}(X \times Y; G_1 \otimes G_2)$$

given by $\{\alpha\} \times \{\beta\} = \{\alpha \times \beta\}$.

PROOF This will follow immediately from three consequences of Proposition 4.6:

   (a)   cocyle $\times$ cocycle is a cocycle;
   (b)   cocycle $\times$ coboundary is a coboundary;
   (c)   coboundary $\times$ cocycle is a coboundary.

If $\delta\alpha = 0 = \delta\beta$, then $\delta(\alpha \times \beta) = (\delta\alpha) \times \beta + (-1)^p \alpha \times \delta\beta = 0$. This establishes (a); (b) and (c) follow in similar fashion. $\square$

The product given by Corollary 4.7 is the *cohomology external product*.

Exercise 4 If $f: X' \to X$ and $g: Y' \to Y$ are maps, $\{\alpha\} \in H^p(X; G_1)$, and $\{\beta\} \in H^q(Y; G_2)$, show that

$$(f \times g)^*(\{\alpha\} \times \{\beta\}) = f^*\{\alpha\} \times g^*\{\beta\}$$

in $H^{p+q}(X' \times Y'; G_1 \otimes G_2)$.

Let R be an associative commutative ring with unit. So there is a homomorphism $\mu: R \otimes R \to R$ given by $\mu(a \otimes b) = ab$. We now specialize the cohomology external product to the case where $G_1 = R = G_2$. For $\alpha \in S^p(X; R)$ and $\beta \in S^q(Y; R)$ define $\alpha \times_1 \beta \in S^{p+q}(X \times Y; R)$ to be the composition

$$S_{p+q}(X \times Y) \xrightarrow{\phi} S_*(X) \otimes S_*(Y) \xrightarrow{\alpha \otimes \beta} R \otimes R \xrightarrow{\mu} R.$$

As before this induces a well-defined product on cohomology groups

$$H^p(X; R) \otimes H^q(Y; R) \to H^{p+q}(X \times Y; R)$$

by taking $\{\alpha\} \otimes \{\beta\}$ into $\{\alpha \times_1 \beta\}$.

4.8 LEMMA Let $\{\alpha\} \in H^p(X; R)$ and $\{\beta\} \in H^q(Y; R)$ and define the map $T: X \times Y \to Y \times X$ by $T(x, y) = (y, x)$. Then

$$T^*: \quad H^{p+q}(Y \times X; R) \to H^{p+q}(X \times Y; R)$$

has

$$T^*(\{\beta\} \times_1 \{\alpha\}) = (-1)^{pq}(\{\alpha\} \times_1 \{\beta\}).$$

PROOF   Define $T': S_*(X) \otimes S_*(Y) \to S_*(Y) \otimes S_*(X)$ on a basis element $e \otimes c$, where $e \in S_p(X)$ and $c \in S_q(Y)$, by

$$T'(e \otimes c) = (-1)^{pq}c \otimes e.$$

Then consider the diagram

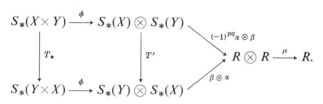

It is evident that $(-1)^{pq}\mu \circ (\alpha \otimes \beta) = \mu \circ (\beta \otimes \alpha) \circ T'$ since R is commutative. Restricting our attention to the rectangle, we observe that the composition $\phi \circ T_*$ is a chain map, since both ϕ and T_* are chain maps. We also claim that $T' \circ \phi$ is a chain map. To establish this it is sufficient to show that T' is a chain map, so let $e \in S_p(X)$ and $c \in S_q(Y)$. Then

$$\begin{aligned}
T' \circ \partial(e \otimes c) &= T'(\partial e \otimes c + (-1)^p e \otimes \partial c) \\
&= (-1)^{(p-1)q}c \otimes \partial e + (-1)^{p+p(q-1)}\partial c \otimes e \\
&= (-1)^{(p-1)q}c \otimes \partial e + (-1)^{pq}\partial c \otimes e.
\end{aligned}$$

On the other hand

$$\begin{aligned}
\partial \circ T'(e \otimes c) &= \partial((-1)^{pq}c \otimes e) \\
&= (-1)^{pq}\partial c \otimes e + (-1)^{pq+q}c \otimes \partial e \\
&= (-1)^{pq}\partial c \otimes e + (-1)^{(p+1)q}c \otimes \partial e.
\end{aligned}$$

Since these two expressions are equal, we conclude that T' is a chain map.

Now if we check on zero-dimensional homology, it is evident that $T' \circ \phi$ and $\phi \circ T_*$ induce the same transformation. By applying the acyclic model theorem, Theorem 4.3, we conclude that these two chain maps are naturally chain homotopic. Therefore, the cohomology class represented by the composition $\mu \circ (\beta \otimes \alpha) \circ \phi \circ T_*$ is the same as the class represented by $(-1)^{pq}\mu \circ (\alpha \otimes \beta) \circ \phi$. In other words

$$T^*(\{\beta\} \times_1 \{\alpha\}) = (-1)^{pq}\{\alpha\} \times_1 \{\beta\}. \qquad \square$$

4.9   LEMMA   If $\{\alpha\} \in H^*(X; R)$ and $\{\beta\} \in H^*(Y; R)$ and $f: X' \to X$, $g: Y' \to Y$ are maps, then

$$(f \times g)^*(\{\alpha\} \times_1 \{\beta\}) = f^*\{\alpha\} \times_1 g^*\{\beta\}.$$

PROOF   This follows routinely as in Exercise 4.   ☐

4.10   LEMMA   If $\{\alpha\} \in H^*(X; R)$, $\{\beta\} \in H^*(Y; R)$, and $\{\gamma\} \in H^*(W; R)$, then

$$(\{\alpha\} \times_1 \{\beta\}) \times_1 \{\gamma\} = \{\alpha\} \times_1 (\{\beta\} \times_1 \{\gamma\}).$$

Exercise 5   Prove Lemma 4.10.   ☐

Observe that if we take $Y = $ point, then $H^*(Y; R) = H^0(Y; R) \approx R$ and the external product

$$H^*(X; R) \otimes H^*(Y; R) \to H^*(X \times Y; R)$$

has the form

$$H^*(X; R) \otimes R \to H^*(X; R).$$

This gives $H^*(X; R)$ the structure of a graded R-module. Moreover, it follows from Lemma 4.9 that any map $f: X' \to X$ induces an R-module homomorphism

$$f^*:\quad H^*(X; R) \to H^*(X'; R).$$

For any space X let $d: X \to X \times X$ be the diagonal mapping given by $d(x) = (x, x)$. Then the composition

$$H^p(X; R) \otimes H^q(X; R) \to H^{p+q}(X \times X; R) \xrightarrow{d^*} H^{p+q}(X; R)$$

sending $\{\alpha\} \otimes \{\beta\}$ into $d^*(\{\alpha\} \times_1 \{\beta\})$ defines a multiplication in the R-module $H^*(X; R)$. This is called the *cup product* and is usually written $\{\alpha\} \cup \{\beta\}$ or just $\{\alpha\} \cdot \{\beta\}$.

By applying the previous lemmas we may conclude the following important result.

4.11   THEOREM   For R a commutative associative ring with unit, X a topological space, $H^*(X; R)$ is a commutative associative graded R-algebra with unit. Any continuous function $f: X' \to X$ induces an R-algebra homomorphism

$$f^*:\quad H^*(X; R) \to H^*(X'; R)\quad \text{of degree zero.}\quad \square$$

As a point of information, a graded R-algebra $M = \sum_k M^k$ is *commutative* if given any homogeneous elements $m_p \in M^p$ and $m_q \in M^q$, we have

$$m_p \cdot m_q = (-1)^{pq} m_q m_p \quad \text{in} \quad M^{p+q}.$$

NOTE It is important to observe that while all of the development of products so far has been in terms of single spaces for the sake of clarity, the same constructions may be duplicated using pairs of spaces and relative homology and cohomology groups. It is important to point out that in this context, the cartesian product of pairs is another pair given by

$$(X, A) \times (Y, B) = (X \times Y, X \times B \cup A \times Y).$$

Exercise 6 Is the connecting homomorphism in the long exact cohomology sequence for the pair (X, A) an R-module homomorphism? An R-algebra homomorphism?

The essential tool used in defining the cup product of two cohomology classes is the composition of chain maps

$$S_*(X) \xrightarrow{d_*} S_*(X \times X) \xrightarrow{\phi} S_*(X) \otimes S_*(X).$$

More generally, suppose that $\tau \colon S_*(X) \to S_*(X) \otimes S_*(X)$ is a chain map such that

(i) $\tau(a) = a \otimes a$ for any singular 0-simplex a;
(ii) τ commutes appropriately with homomorphisms induced by maps of spaces.

Then by applying the acyclic model theorem, Theorem 4.3, we see that any such τ must be chain homotopic to $\phi \circ d_*$. This implies that the cup product on cohomology classes is independent of the choice of τ as long as the stated conditions are satisfied. A chain map τ with these properties is usually called a *diagonal approximation*. For use in later definitions and examples it will be helpful to have a specific example for τ. The following is the *Alexander–Whitney* diagonal approximation.

Given a singular n-simplex $\phi \colon \sigma^n \to X$ in a space X, define the *front i-face* $_i\phi$, $0 \leq i \leq n$, to be the singular i-simplex

$$_i\phi(t_0, \ldots, t_i) = \phi(t_0, \ldots, t_i, 0, \ldots, 0).$$

Similarly let the *back j-face* ϕ_j, $0 \leq j \leq n$, be the singular j-simplex

$$\phi_j(t_0, \ldots, t_j) = \phi(0, \ldots, 0, t_0, \ldots, t_j).$$

Then define

$$\tau(\phi) = \sum_{i+j=n} {}_i\phi \otimes \phi_j$$

for ϕ a singular n-simplex in X.

For example, if $\phi: \sigma^2 \to \sigma^2$ is the identity, then $\tau(\phi) = 0 \otimes \phi + (0, 1) \otimes (1, 2) + \phi \otimes 2$ where 0 and 2 are the obvious 0-simplices and $(0, 1)$ and $(1, 2)$ are 1-simplices (see Figure 4.1).

It is evident that Properties (i) and (ii) above are satisfied by τ. The only mild complication is left as the following exercise.

Exercise 7 Show that the Alexander–Whitney diagonal approximation τ is a chain map.

Using this specific model for τ, let us see exactly what the cup product looks like. Let $\alpha \in S^p(X; R)$ and $\beta \in S^q(X; R)$ and ϕ be a singular $(p + q)$-simplex in X. Then

$$\langle \alpha \cup \beta, \phi \rangle$$

is the image of the composition

$$\phi \xrightarrow{\tau} \sum {}_i\phi \otimes \phi_j \xrightarrow{\alpha \otimes \beta} \alpha({}_p\phi) \otimes \beta(\phi_q) \xrightarrow{\mu} \alpha({}_p\phi) \cdot \beta(\phi_q).$$

Thus, $\langle \alpha \cup \beta, \phi \rangle = \langle \alpha, {}_p\phi \rangle \cdot \langle \beta, \phi_q \rangle$.

Example We want to compute the cohomology ring of the two-dimensional torus $T^2 = S^1 \times S^1$. Recall that $H_1(T^2; Z) \approx Z \oplus Z$, and the generators may

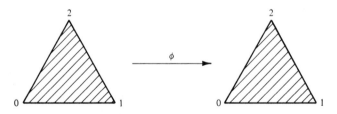

Figure 4.1

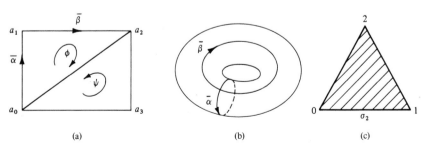

(a)　　　　　(b)　　　　　(c)

Figure 4.2

be represented by $\bar{\alpha}$ and $\bar{\beta}$ in Figure 4.2b. For $H_2(T^2; Z) \approx Z$ we may use as generator the 2-chain $\phi - \psi$, where (Figure 4.2c)

$$\phi(0) = a_0, \qquad \phi(1) = a_1, \qquad \phi(2) = a_2$$

and

$$\psi(0) = a_0, \qquad \psi(1) = a_3, \qquad \psi(2) = a_2$$

(see Figure 4.2a).

Using the universal coefficient theorem, Theorem 3.14, we see that $H^1(T^2; Z) \approx \text{Hom}(H_1(T^2; Z), Z)$ and we choose as generators α, β, where

$$\alpha(\bar{\alpha}) = 1, \qquad \alpha(\bar{\beta}) = 0,$$
$$\beta(\bar{\alpha}) = 0, \qquad \beta(\bar{\beta}) = 1.$$

Now

$$\langle \alpha \cup \beta, \phi - \psi \rangle = \langle \alpha, {}_1\phi \rangle \cdot \langle \beta, \phi_1 \rangle - \langle \alpha, {}_1\psi \rangle \cdot \langle \beta, \psi_1 \rangle$$
$$= \langle \alpha, \bar{\alpha} \rangle \cdot \langle \beta, \bar{\beta} \rangle - \langle \alpha, \bar{\beta} \rangle \cdot \langle \beta, \bar{\alpha} \rangle$$
$$= 1 - 0$$
$$= 1.$$

On the other hand

$$\langle \alpha \cup \alpha, \phi - \psi \rangle = \langle \alpha, \bar{\alpha} \rangle \cdot \langle \alpha, \bar{\beta} \rangle - \langle \alpha, \bar{\beta} \rangle \cdot \langle \alpha, \bar{\alpha} \rangle$$
$$= 0.$$

Similarly $\langle \beta \cup \beta, \phi - \psi \rangle = 0$. Since $H^2(T^2; Z) \approx \text{Hom}(H_2(T^2; Z), Z) \approx Z$, we now have computed the cohomology ring $H^*(T^2; Z)$. Thus, $H^*(T^2; Z)$ is the graded algebra over Z generated by elements α and β of degree 1, subject to the relations

$$\alpha^2 = 0, \qquad \beta^2 = 0, \qquad \alpha\beta = -\beta\alpha.$$

NOTE   This has the form of an exterior algebra on two generators. How about the cohomology ring of the n-torus $T^n = S^1 \times \cdots \times S^1$?

Suppose that $f: S^{2n-1} \to S^n$ is a map, $n \geq 2$. There is a procedure for associating with such a map an integer $H(f)$, the *Hopf invariant* of f. This may be defined using the cup product in the following way: Let $\{\alpha\}$ and $\{\beta\}$ be generators of the cohomology groups $H^{2n-1}(S^{2n-1}; Z)$ and $H^n(S^n; Z)$, respectively, represented by the cocycles α and β. Since $\{\beta\} \cup \{\beta\} = 0$, the cocycle $\beta \cup \beta$ must be a coboundary. That is, there exists a cochain

$\gamma \in S^{2n-1}(S^n; Z)$ with

$$\delta\gamma = \beta \cup \beta.$$

Since $H^n(S^{2n-1}; Z) = 0$, the cocycle $f^{\#}(\beta) \in S^n(S^{2n-1}; Z)$ must be a coboundary, and there exists a cochain ε in $S^{n-1}(S^{2n-1}; Z)$ such that $\delta\varepsilon = f^{\#}(\beta)$. Now $\varepsilon \cup f^{\#}(\beta)$ and $f^{\#}(\gamma)$ are cochains in $S^{2n-1}(S^{2n-1}; Z)$. Moreover

$$\delta(\varepsilon \cup f^{\#}(\beta) - f^{\#}(\gamma)) = \delta(\varepsilon \cup \delta\varepsilon) - f^{\#}(\beta \cup \beta)$$
$$= \delta\varepsilon \cup \delta\varepsilon - f^{\#}(\beta) \cup f^{\#}(\beta)$$
$$= 0.$$

So we define $H(f)$ to be the integer which when multiplied times $\{\alpha\}$ gives the cohomology class of $\varepsilon \cup f^{\#}(\beta) - f^{\#}(\gamma)$. That is

$$\{\varepsilon \cup f^{\#}(\beta) - f^{\#}(\gamma)\} = H(f) \cdot \{\alpha\}.$$

Exercise 8 (a) Let $f: S^{2n-1} \to S^n$ be a map of Hopf invariant k. If $\bar{g}: S^{2n-1} \to S^{2n-1}$ and $g: S^n \to S^n$ are maps of degree p, determine $H(gf)$ and $H(f\bar{g})$.
(b) If

$$\begin{array}{ccc} S^{2n-1} & \xrightarrow{\bar{h}} & S^{2n-1} \\ \downarrow{\scriptstyle f} & & \downarrow{\scriptstyle f} \\ S^n & \xrightarrow{h} & S^n \end{array}$$

is a commutative diagram where f has Hopf invariant k, how are the degrees of h and $\bar{h}$ related?

Let us give an alternate definition for $H(f)$. Recall that S^n and S^{2n-1} may be given the structure of finite CW complexes, each having only two cells. Given a map $f: S^{2n-1} \to S^n$, we denote by S_f^n the space obtained by attaching a $2n$-cell to S^n via f (see Chapter 2). Then S_f^n is a finite CW complex with three cells, of dimension 0, n, and $2n$. Applying the technique of Theorem 2.21 we see that since $n > 1$, the cohomology of S_f^n is given by

$$H^i(S_f^n) \approx \begin{cases} Z & \text{for } i = 0, n, 2n \\ 0 & \text{otherwise.} \end{cases}$$

Denoting by $b \in H^n(S_f^n; Z)$ and $a \in H^{2n}(S_f^n; Z)$ a chosen pair of generators, we define $H(f)$ to be that integer for which $b^2 = H(f) \cdot a$ in $H^{2n}(S_f^n; Z)$.

Exercise 9 Show that the two definitions of $H(f)$ are equivalent.

In order to show that $H(f)$ is an invariant of the homotopy class of f, we need the following result due to J. H. C. Whitehead.

4.12 PROPOSITION If $f_0, f_1: S^p \to X$ are homotopic maps into a space X, then the identity map of X extends to a homotopy equivalence

$$h: \quad X_{f_0} \to X_{f_1}.$$

PROOF Let $\{f_t\}$ be a homotopy between f_0 and f_1 and denote an element of D^{p+1} by θu, where $u \in S^p$ and $0 \le \theta \le 1$.

Given a radius in the attached disk in X_{f_0} (Figure 4.3a), the inner half should be mapped onto the corresponding radius in X_{f_1}. Then the outer half is used to trace out the path of the homotopy from $f_1(u)$ to $f_0(u)$ (Figure 4.3b). Specifically then define the map h by

$$h(x) = x \qquad \text{for} \quad x \in X;$$
$$h(\theta u) = 2\theta u \qquad \text{for} \quad u \in S^p, \quad 0 \le \theta \le \tfrac{1}{2};$$
$$h(\theta u) = f_{2-2\theta}(u) \qquad \text{for} \quad u \in S^p, \quad \tfrac{1}{2} \le \theta \le 1.$$

Defining a similar map $h': X_{f_1} \to X_{f_0}$, it is easily seen that the compositions $h \circ h'$ and $h' \circ h$ are homotopic to the respective identities. $\square$

4.13 PROPOSITION If $f_0, f_1: S^{2n-1} \to S^n$ are homotopic maps, then $H(f_0) = H(f_1)$.

PROOF Let $h: S^n_{f_0} \to S^n_{f_1}$ be the homotopy equivalence given in Proposition 4.12. If $i_0: (D^{2n}, S^{2n-1}) \to (S^n_{f_0}, S^n)$ and $i_1: (D^{2n}, S^{2n-1}) \to (S^n_{f_1}, S^n)$,

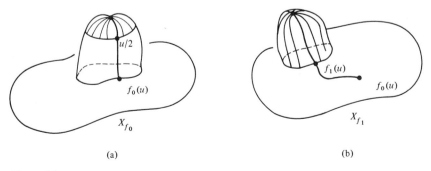

(a)   (b)

Figure 4.3

denote the relative homeomorphisms corresponding to f_0 and f_1, the diagram

$$(D^{2n}, S^{2n-1}) \xrightarrow{\ i_0\ } (S^n_{f_0}, S^n)$$

with i_1 going down-left and h going down-right to

$$(S^n_{f_1}, S^n)$$

is homotopy commutative. This homotopy is easily defined by setting $g_t(\theta u) = h \circ i_0((1 - \tfrac{1}{2}t)\theta u)$.

This implies that the diagram of cohomology groups

$$H^{2n}(S^n_{f_1}, S^n) \xrightarrow{\ h^*\ } H^{2n}(S^n_{f_0}, S^n)$$

with i_1^* and i_0^* going to

$$H^{2n}(D^{2n}, S^{2n-1})$$

is commutative. Thus, a choice of an orientation for D^{2n} dictates compatible choices of generators

$$\bar{a}_1 \in H^{2n}(S^n_{f_1}, S^n) \qquad \text{and} \qquad \bar{a}_0 \in H^{2n}(S^n_{f_0}, S^n)$$

and corresponding choices of generators

$$a_1 \in H^{2n}(S^n_{f_1}) \qquad \text{and} \qquad a_0 \in H^{2n}(S^n_{f_0})$$

such that $h^*(a_1) = a_0$.

Furthermore, if $b_1 \in H^n(S^n_{f_1})$ and $b_0 \in H^n(S^n_{f_0})$ are generators corresponding to a chosen orientation of D^n, then since h is the identity on S^n, it follows that $h^*(b_1) = b_0$.

Therefore

$$H(f_0) \cdot a_0 = b_0^2 = (h^*(b_1))^2$$
$$= h^*(b_1^2)$$
$$= h^*(H(f_1) \cdot a_1)$$
$$= H(f_1) \cdot h^*(a_1)$$
$$= H(f_1) \cdot a_0,$$

and

$$H(f_0) = H(f_1). \qquad \square$$

NOTE If n is odd, the commutativity of the cup product implies that $b^2 = -b^2$, so that $H(f) = 0$. Thus, the Hopf invariant can only be non-zero for maps $f: S^{4n-1} \to S^{2n}$.

4.14 PROPOSITION For any $n > 0$ there exist maps from S^{4n-1} to S^{2n} of arbitrary even Hopf invariant.

PROOF As a corollary of Exercise 8, it is sufficient to show that there exists a map with Hopf invariant ± 2.

Recall that $S^{2n} \times S^{2n}$ may be given the structure of a finite CW complex having one 0-cell, two $2n$-cells, and one $4n$-cell (see Proposition 2.6). Furthermore, there is a map

$$f: \quad S^{4n-1} \to S^{2n} \vee S^{2n}$$

where $S^{2n} \vee S^{2n}$ is the $2n$-skeleton of $S^{2n} \times S^{2n}$, such that $S^{2n} \times S^{2n}$ is the space obtained by attaching a $4n$-cell to $S^{2n} \vee S^{2n}$ via f.

Define a map

$$g: \quad S^{2n} \vee S^{2n} \to S^{2n}$$

by $g(x, p) = x$, $g(p, y) = y$, where $S^{2n} \vee S^{2n}$ is identified with

$$(S^{2n} \times p) \cup (p \times S^{2n}) \subseteq S^{2n} \times S^{2n}.$$

From the commutative diagram

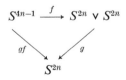

we see that g induces a map

$$\tilde{g}: \quad S^{2n} \times S^{2n} = (S^{2n} \vee S^{2n})_f \to S^{2n}_{gf}.$$

Using this map $\tilde{g}$ we want to prove that the Hopf invariant of gf is ± 2.

Let $e \in H^0(p)$, $1 \in H^0(S^{2n})$, and $c \in H^{2n}(S^{2n})$ be generators of these infinite cyclic groups. Then $H^{4n}(S^{2n} \times S^{2n})$ is the infinite cyclic group generated by $c \times c$ and $H^{2n}(S^{2n} \times S^{2n})$ is the free abelian group with basis consisting of $1 \times c$ and $c \times 1$. As before let $a \in H^{4n}(S^{2n}_{gf})$ and $b \in H^{2n}(S^{2n}_{gf})$ be generators of these infinite cyclic groups.

First, we must compute $\tilde{g}^*(b) \in H^{2n}(S^{2n} \times S^{2n})$. If $j: p \to S^{2n}$ is the inclusion, then both rectangles in the following diagram commute:

$$
\begin{array}{ccccc}
H^{2n}(S^{2n}_{gf}) & \xrightarrow{\tilde{g}*} & H^{2n}(S^{2n}\times S^{2n}) & \xleftarrow{\times} & H^{2n}(S^{2n})\otimes H^0(S^{2n}) \\
\approx\downarrow i* & & \downarrow (\mathrm{id}\times j)* & & \approx\downarrow (\mathrm{id})*\otimes j* \\
H^{2n}(S^{2n}) & \xrightarrow{=} & H^{2n}(S^{2n}\times p) & \xleftarrow[\approx]{\times} & H^{2n}(S^{2n})\otimes H^0(p)
\end{array}
$$

Thus

$$
i^*(b) = \pm c\times e = \pm(\mathrm{id})^*(c)\times j^*(1)
$$

or

$$
(\mathrm{id}\times j)^*\tilde{g}^*(b) = \pm(\mathrm{id}\times j)^*(c\times 1).
$$

This means that the element

$$
\tilde{g}^*(b) \pm c \times 1
$$

is in the kernel of $(\mathrm{id}\times j)^*$ for some choice of sign. Now the kernel of $(\mathrm{id}\times j)^*$ in $H^{2n}(S^{2n}\times S^{2n})$ is the infinite cyclic subgroup generated by $1\times c$, so

$$
\tilde{g}^*(b) \pm c\times 1 = m(1\times c)
$$

for some integer m.

By the same argument, $\tilde{g}^*(b) \pm 1\times c = k(c\times 1)$ for some integer k. These two properties together imply that with a proper choice of sign, we have

$$
\tilde{g}^*(b) = \pm c\times 1 \pm 1\times c.
$$

It can be easily checked that

$$
\begin{aligned}
(\pm c\times 1 \pm 1\times c)^2 &= (c\times 1)^2 \pm 2(c\times 1)\cdot(1\times c) + (1\times c)^2 \\
&= c^2\times 1 \pm 2c\times c + 1\times c^2 \\
&= \pm 2c\times c,
\end{aligned}
$$

since $c^2 = 0$.

Finally, since

$$
\tilde{g}^*: \quad H^{4n}(S^{2n}_{gf}) \to H^{4n}(S^{2n}\times S^{2n})
$$

is an isomorphism,

$$
\begin{aligned}
b^2 &= \tilde{g}^{*-1}(\tilde{g}^*(b^2)) = \tilde{g}^{*-1}((\pm c\times 1 \pm 1\times c)^2) \\
&= \tilde{g}^{*-1}(\pm 2c\times c) \\
&= \pm 2a.
\end{aligned}
$$

This proves that $H(gf) = \pm 2$. $\square$

There remains the question of the existence of maps having odd Hopf invariant. Using the results of the next chapter we will be able to show that the Hopf maps $S^3 \to S^2$ and $S^7 \to S^4$ each have Hopf invariant one. By using the Cayley numbers, one can define an analogous map $S^{15} \to S^8$ of Hopf invariant one. Results of Adem [1952] on certain cohomology operations imply that there can exist maps $f: S^{4n-1} \to S^{2n}$ of odd Hopf invariant only when n is a power of 2. Finally, there is a deep theorem due to Adams [1960], that for $n \neq 1, 2$, or 4 there is no map $f: S^{4n-1} \to S^{2n}$ of odd Hopf invariant. An important consequence of this theorem is that the only values of n for which $\mathbb{R}^n$ carries the structure of a real division algebra are: $n = 1$ (real numbers), $n = 2$ (complex numbers), and $n = 4$ (quaternions). [See Eilenberg and Steenrod, 1952, p. 320].

As a reference for further information on the Hopf invariant we recommend Hu [1959]. We cite only briefly one further result: two maps from S^3 to S^2 are homotopic if and only if they have the same Hopf invariant.

As the final topic of this chapter we introduce a variant of the cup product which will be useful in the following chapter. Let X be a space and R be a commutative ring with unit. If $\alpha \in S^p(X; R)$ we may view α as a homomorphism of all $S_*(X)$ into R by setting it equal to zero on elements of dimension different from p.

The composition

$$S_*(X) \xrightarrow{\tau} S_*(X) \otimes S_*(X) \xrightarrow{\alpha \otimes \mathrm{id}} R \otimes S_*(X)$$

when tensored throughout with R yields

$$R \otimes S_*(X) \xrightarrow{\mathrm{id} \otimes \tau} R \otimes S_*(X) \otimes S_*(X) \xrightarrow{\mathrm{id} \otimes \alpha \otimes \mathrm{id}} R \otimes R \otimes S_*(X)$$
$$\xrightarrow{\mu \otimes \mathrm{id}} R \otimes S_*(X).$$

If $c \in S_n(X; R) = R \otimes S_n(X)$, we define the *cap product* of α and c, $\alpha \cap c$, to be the image of c under this composition. Note that

$$\alpha \cap c \in S_{n-p}(X; R).$$

For example, suppose τ is the Alexander–Whitney diagonal and ϕ is a singular n-simplex. Then the above composition has

$$1 \otimes \phi \to 1 \otimes \{ \sum_{i+j=n} {}_i\phi \otimes \phi_j \} \to 1 \otimes \alpha({}_p\phi) \otimes \phi_{n-p} \to \alpha({}_p\phi) \otimes \phi_{n-p}.$$

If we interpret $R \otimes S_*(X)$ as the free R-module generated by the singular simplices of X, then

$$\alpha \cap \phi = \alpha(_p\phi) \cdot \phi_{n-p}.$$

It is evident that this is closely connected to the cup product. To make this relationship specific, let $\alpha \in S^p(X; R)$, $\beta \in S^q(X; R)$ and $\phi \in S_{p+q}(X; R)$, where ϕ is a singular simplex. Then

$$\langle \beta \cup \alpha, \phi \rangle = \beta(_q\phi) \cdot \alpha(\phi_p).$$

On the other hand

$$\langle \alpha, \beta \cap \phi \rangle = \langle \alpha, \beta(_q\phi) \cdot \phi_p \rangle = \beta(_q\phi) \cdot \alpha(\phi_p).$$

Since this is true for all ϕ, it follows that for any $c \in S_{p+q}(X; R)$,

(4.15) $$\langle \beta \cup \alpha, c \rangle = \langle \alpha, \beta \cap c \rangle.$$

Finally, we must determine the action of ∂ on the chain $\alpha \cap c$. To do this we evaluate an arbitrary cochain γ on $\partial(\alpha \cap c)$,

$$\langle \gamma, \partial(\alpha \cap c) \rangle = \langle \delta\gamma, \alpha \cap c \rangle = \langle \alpha \cup \delta\gamma, c \rangle.$$

Suppose that $\alpha \in S^q(X; R)$, $c \in S_n(X; R)$ so that $\alpha \cap c \in S_{n-q}(X; R)$. Recall that

$$\delta(\alpha \cup \gamma) = \delta\alpha \cup \gamma + (-1)^q\alpha \cup \delta\gamma$$

or

$$\alpha \cup \delta\gamma = (-1)^q(\delta(\alpha \cup \gamma) - (\delta\alpha) \cup \gamma).$$

So by substituting into the previous equation we have

$$\langle \gamma, \partial(\alpha \cap c) \rangle = (-1)^q \langle \delta(\alpha \cup \gamma - \delta\alpha \cup \gamma, c \rangle$$
$$= (-1)^q[\langle \gamma, \alpha \cap \delta c \rangle - \langle \gamma, \delta\alpha \cap c \rangle]$$
$$= \langle \gamma, (-1)^q(\alpha \cap \partial c - \delta\alpha \cap c) \rangle.$$

Since this is true for all cochains γ, it follows that

$$(-1)^q\partial(\alpha \cap c) = (\alpha \cap \partial c) - (\delta\alpha \cap c).$$

From this derivation formula we conclude that the cap product on chain groups induces a well-defined product on homology groups which takes the form

$$H^q(X; R) \otimes H_n(X; R) \to H_{n-q}(X; R)$$

and sends $\{\alpha\} \otimes \{c\}$ into $\{\alpha \cap c\}$.

Exercise 10   Formulate and prove a statement showing that the cap product is natural with respect to homomorphisms induced by mappings of spaces.

Exercise 11   A graded group $\{G_q\}$ is said to be of *finite type* if for each q, G_q is finitely generated. Prove the following theorem.

4.16 THEOREM (*Cohomology Künneth formula*)   Let G and G' be abelian groups with $\mathrm{Tor}(G, G') = 0$. If $H_*(X; Z)$ and $H_*(Y; Z)$ are of finite type, then there is a split exact sequence

$$0 \to \sum_{p+q=n} H^p(X; G) \otimes H^q(Y; G') \xrightarrow{\ \times\ } H^n(X \times Y; G \otimes G')$$

$$\to \sum_{p+q=n+1} \mathrm{Tor}(H^p(X; G), H^q(Y; G')) \to 0. \qquad \square$$

Exercise 12   Let X and Y be spaces and R be a commutative ring. If $u_1 \in H^p(X; R)$, $u_2 \in H^q(X; R)$, $v_1 \in H^r(Y; R)$, and $v_2 \in H^s(Y; R)$, then in $H^{p+q+r+s}(X \times Y; R)$ we have

$$(u_1 \times v_1) \cup (u_2 \times v_2) = (-1)^{qr}(u_1 \cup u_2) \times (v_1 \cup v_2).$$

Exercise 13   If $u \in H^p(X; R)$, $v \in H^q(Y; R)$, $p_1: X \times Y \to X$,   and $p_2: X \times Y \to Y$ are the projection maps, then

$$u \times v = p_1^*(u) \cup p_2^*(v).$$

Exercise 14   If $u_1 \in H^p(X; R)$, $u_2 \in H^q(Y; R)$, $z_1 \in H_m(X; R)$,   and $z_2 \in H_n(Y; R)$, then in $H_{m+n-p-q}(X \times Y; R)$ we have

$$(u_1 \times u_2) \cap (z_1 \times z_2) = (-1)^{q(m-p)}(u_1 \cap z_1) \times (u_2 \cap z_2).$$

Exercise 15   Show that the cap product may be extended to relative homology and cohomology groups of a pair to give products of the form

$$H^k(X, A) \otimes H_n(X, A) \to H_{n-k}(X)$$

and

$$H^k(X) \otimes H_n(X, A) \to H_{n-k}(X, A).$$

chapter 5

MANIFOLDS AND
POINCARÉ DUALITY

This chapter deals with some of the basic homological properties of topological manifolds. Since the main result is the Poincaré duality theorem, we begin with a simple example to establish an intuitive feeling for this classical result. This is followed by material on topological manifolds and a detailed proof of the theorem. The approach used follows the excellent treatment of Samelson [1965, pp. 323–336] and proceeds by way of the Thom isomorphism theorem. Several applications of the theorem follow, including the determination of the cohomology rings of projective spaces and results on the index of topological manifolds and cobordism.

Before proceeding with the general approach, let us see how the theorem may be motivated from an example. Briefly, the Poincaré duality theorem will say that if M is a compact oriented n-manifold without boundary, the ith Betti number of M is the same as the $(n-i)$th Betti number for $0 \leq i \leq n$. In the following example we will indicate how such a correspondence arises.

Suppose we are given a portion of a triangulated surface K as shown in Figure 5.1. By taking the first barycentric subdivision (see Appendix I) we arrive at a new triangulation K', as shown in Figure 5.2. If v is a vertex in this new triangulation, define the *star* of v in K' to be the union of all open cells in K' that contain v in their closure. Thus, star$(A; K')$ is the open 2-cell shown in Figure 5.3, whereas star(v_0; K') is the open 2-cell shown in Figure 5.4.

Given a simplex σ in K, we define its *dual cell* σ^* in K' by

$$\sigma^* = \bigcap_v \overline{\text{star}}(v; K'),$$

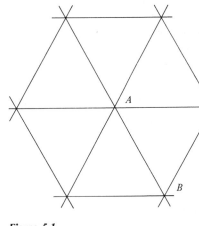

Figure 5.1

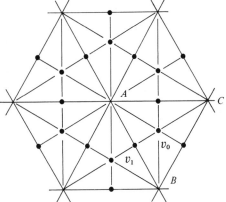

Figure 5.2

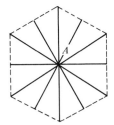

Figure 5.3

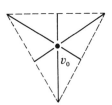

Figure 5.4

where v ranges over the vertices of σ. For example, the dual of the vertex A is $\overline{\text{star}}(A; K')$, the closure of Figure 5.3, while the dual of the 2-simplex ABC is the vertex v_0. Similarly the dual of the 1-simplex AB is the 1-cell joining v_0 and v_1. Note that while the dual of a simplex need not be a simplex, it is a cell in the complementary dimension.

As a specific example we take the boundary of a 3-simplex, a triangulated surface homeomorphic to S^2 (Figure 5.5). This surface may be viewed as a finite CW complex having four 0-cells, six 1-cells, and four 2-cells. By taking the dual cells of each of these simplices we get a corresponding CW decomposition for the same space as shown in Figure 5.6. Here we have four 2-cells (A^*, B^*, C^*, D^*), six 1-cells ($AB^*, \ldots, CD^*$), and four 0-cells ($ABC^*, \ldots, BCD^*$).

To compute the Betti numbers of these complexes we use the cellular chain complex of Theorem 2.21. Recall that if Y is a finite CW complex and

$$C_{i+1}(Y) \xrightarrow{\partial_{i+1}} C_i(Y) \xrightarrow{\partial_i} C_{i-1}(Y)$$

is a portion of the chain complex, then the ith Betti number

$$\beta_i(Y) = \alpha_i(Y) - \gamma_{i+1}(Y) - \gamma_i(Y),$$

where $\alpha_i(Y)$ is the number of i-cells in Y and $\gamma_j(Y)$ is the rank of the image of ∂_j.

Denoting by X and X^* the two structures above we may make the following comparisons:

$$\alpha_0(X) = 4 \qquad\qquad\qquad \alpha_2(X^*) = 4$$

given by A, B, C, D \qquad given by A^*, B^*, C^*, D^*

$$\alpha_1(X) = 6 \qquad\qquad\qquad \alpha_1(X^*) = 6$$

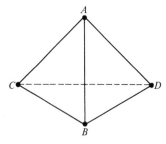

Figure 5.5

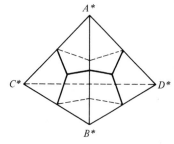

Figure 5.6

given by $AB, \ldots, CD$ given by $AB^*, \ldots, CD^*$

$$\alpha_2(X) = 4 \qquad\qquad \alpha_0(X^*) = 4$$

given by $ABC, \ldots, BCD$ given by $ABC^*, \ldots, BCD^*$

$$\gamma_0(X) = 0 \qquad\qquad \gamma_3(X^*) = 0$$
$$\gamma_1(X) = 3, \text{ basis} \qquad \gamma_2(X^*) = 3, \text{ basis}$$

given by $A - B,\ A - C,\ A - D$ given by $\partial(A^* - B^*),$
$$\partial(A^* - C^*),\ \partial(A^* - D^*)$$

$$\gamma_2(X) = 3, \text{ basis} \qquad \gamma_1(X^*) = 3, \text{ basis}$$

given by $\partial(ABC - ABD),$ given by $ABC^* - ABD^*,$

$$\partial(ABC - ADC),\ \partial(ABC - BCD) \qquad ABC^* - ADC^*,\ ABC^* - BCD^*$$
$$\gamma_3(X) = 0. \qquad\qquad \gamma_0(X^*) = 0.$$

Putting this information together, it is evident that $\beta_0(X) = \beta_2(X^*) = \beta_1(X) = \beta_1(X^*) = 0$, and $\beta_2(X) = \beta_0(X^*) = 1$. It may be helpful to keep this sort of geometric picture in mind as we develop the algebraic techniques necessary to establish the theorem in its general setting.

In R^n define the half-space H^n to be the set of all points $(x_1, \ldots, x_n)$ such that $x_n \geq 0$. A *topological n-manifold* is a Hausdorff space M having a countable basis of open sets, with the property that every point of M has a neighborhood homeomorphic to an open subset of H^n. The *boundary* of M, denoted ∂M, is the set of all points x in M for which there exists a homeomorphism of some neighborhood of x onto an open set in H^n taking x into $\{(x_1, \ldots, x_n) \mid x_n = 0\} = \partial H^n \subseteq H^n$.

Exercise 1 Let h be a homeomorphism of an open subset U of H^n onto an open subset of H^n. If $x \in U \cap \partial H^n$, then show $h(x) \in \partial H^n$.

It follows immediately from this exercise that if $x \in \partial M$, then *all* homeomorphisms from open sets about x to open sets in H^n must map x into ∂H^n. M is an *n-manifold without boundary* if $\partial M = \varnothing$, or, equivalently, if each $x \in M$ has a neighborhood homeomorphic to an open set in R^n. A *closed n-manifold* is a compact n-manifold without boundary.

Examples (1) Any open subset of H^n is obviously an n-manifold.

(2) For each point $x \in S^n$, stereographic projection from $-x$ is a homeomorphism from $S^n - \{-x\}$ onto R^n. This gives S^n the structure of a closed n-manifold.

(3) For each point $y \in RP(n)$ pick a point $x \in S^n$ with $\pi(x) = y$, where

$$\pi: \quad S^n \to RP(n)$$

is the identification map. Let $i: (D^n - S^{n-1}) \to S^n$ be the inclusion of the open hemisphere centered at x. Then $\pi \circ i$ is a homeomorphism of an open subset of R^n onto an open set about y. Therefore, $RP(n)$ is a closed n-manifold.

(4) Let $GL(n)$ denote the set of all real $n \times n$ matrices having nonzero determinant. By ordering the entries we may view $GL(n)$ as a subspace of R^{n^2} and give it the induced topology. Under this identification, the determinant function $R^{n^2} \to R$ is continuous and has $GL(n)$ as the inverse of the open set $R - \{0\}$. Thus, $GL(n)$ is an open subset of R^{n^2} and hence is an n^2-manifold without boundary.

(5) The Möbius band, formed by identifying the two ends of a rectangle so that the indicated arrows coincide (Figure 5.7), is obviously a 2-manifold with boundary.

5.1 LEMMA If U is an open subset of R^n, then $H_i(U) = 0$ for $i \geq n$.

PROOF Before proceeding with the proof, we point out a slight generalization of the chain complex in Theorem 2.21. If (X, A) is a finite CW pair, the groups

$$H_p(X^p \cup A, X^{p-1} \cup A)$$

form a chain complex whose homology is $H_*(X, A)$. Note that if every cell of X of dimension greater than p is contained in A, then $H_i(X, A) = 0$ for $i > p$.

Figure 5.7

Let $\{z\} \in H_i(U)$ be an homology class represented by an i-cycle z, $i \geq n$. The image of each singular simplex in U is a compact subset. Since z is a finite linear combination of singular i-simplices, the union of the associated images forms a compact subset $X \subseteq U$.

Define

$$\varepsilon = \mathrm{Inf}\{\| x - y \| \mid x \in X, y \in R^n - U\}.$$

Note that $\varepsilon > 0$, since X is compact, $R^n - U$ is closed, and $X \cap (R^n - U) = \varnothing$. Since X is compact, there exists a large simplex $\mathbb{S}^n$ in R^n such that X is contained in the interior of $\mathbb{S}^n$.

From Appendix I we know that there exists an integer m with mesh $\mathrm{Sd}^m\mathbb{S}^n < \varepsilon$. Now consider $\mathrm{Sd}^m\mathbb{S}^n$ as a finite CW complex under the simplicial decomposition. Let K be the subcomplex of $\mathrm{Sd}^m\mathbb{S}^n$ consisting of all faces of simplices which intersect X (Figure 5.8). Note that by construction

$$X \subseteq K \subseteq U.$$

A portion of the exact homology sequence of the finite CW pair $(\mathrm{Sd}^m\mathbb{S}^n, K)$ has the form

$$\cdots \to H_{i+1}(\mathrm{Sd}^m\mathbb{S}^n, K) \to H_i(K) \to H_i(\mathrm{Sd}^m\mathbb{S}^n) \to \cdots.$$

By our previous comments, $H_{i+1}(\mathrm{Sd}^m\mathbb{S}^n, K) = 0$. Also $H_i(\mathrm{Sd}^m\mathbb{S}^n) = 0$ since the space is a simplex, hence, a convex subset of R^n (see Theorem 1.8). Therefore, $H_i(K) = 0$.

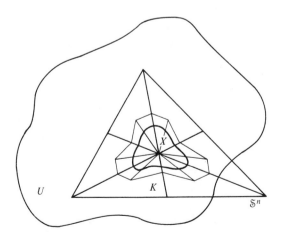

Figure 5.8

Since z was a cycle in X, it is also a cycle in K. The fact that $H_i(K) = 0$ implies that z bounds an $(i + 1)$-chain in K. But this $(i + 1)$-chain also lies in U; hence, z bounds a chain in U and $\{z\} = 0$. $\square$

5.2 LEMMA If M is an n-manifold without boundary, then $H_i(M) = 0$ for $i > n$.

PROOF Let z be an i-cycle in M, $i > n$. Then as in Lemma 5.1 we associate with z the compact subset $X \subseteq M$ which is the union of the images of the singular simplices which make up z. There exists a finite collection $U_1, \ldots, U_m$ of open sets in M, each homeomorphic to an open set in R^n, with $X \subseteq \bigcup U_j$. (Open sets of this type will usually be called *coordinate neighborhoods* or *coordinate charts*.) We will show that $\{z\} = 0$ by proving that $H_i(\bigcup U_j) = 0$ so that z must bound in $\bigcup U_j$.

Proceeding by induction on the number of coordinate neighborhoods, it is true for $m = 1$ by Lemma 5.1. Suppose that

$$H_i\left(\bigcup_{j=1}^{r} U_j \right) = 0.$$

There is a Mayer–Vietoris sequence for the space $(\bigcup_{j=1}^{r} U_j) \cup U_{r+1}$, which has the form

$$\cdots \to H_i\left(\bigcup_{j=1}^{r} U_j \right) \oplus H_i(U_{r+1}) \to H_i\left(\bigcup_{j=1}^{r+1} U_j \right) \to H_{i-1}\left(\left(\bigcup_{j=1}^{r} U_j \right) \cap U_{r+1} \right) \to \cdots.$$

The term on the right is zero by Lemma 5.1 since the space is an open subset of a coordinate chart; similarly $H_i(U_{r+1}) = 0$. It follows from the inductive hypothesis that $H_i(\bigcup_{j=1}^{i+1} U_j) = 0$. This completes the inductive step. $\square$

This result tells us that the nontrivial homology of such a manifold all occurs in dimensions less than or equal to the dimension of the manifold. When the manifold is connected but not compact, this result may be refined to show that the top-dimensional homology group (dimension of the manifold) must also be zero. To establish this, we need the following lemmas.

5.3 LEMMA Let U be open in R^n and $a \in H_n(R^n, U)$. If for every $p \in R^n - U$, the homomorphism induced by inclusion

$$j_{p*} \colon \quad H_n(R^n, U) \to H_n(R^n, R^n - p)$$

has $j_{p*}(a) = 0$, then $a = 0$.

PROOF   The connecting homomorphism for the exact sequence of the pair (R^n, U) gives an isomorphism

$$\Delta: \quad H_n(R^n, U) \xrightarrow{\ \approx\ } H_{n-1}(U).$$

We will prove that $a = 0$ by establishing that $\Delta(a) = 0$ in $H_{n-1}(U)$.

So let $b = \Delta(a)$. Once again, since the "image" of a cycle representing b is a compact subset of U, there exists an open set V with $\bar{V}$ compact $\subseteq U$ and an element b' in $H_{n-1}(V)$ with $i_*(b') = b$, $i: V \to U$ the inclusion.

Let Q be an open cube containing V and define $K = Q - Q \cap U$ (Figure 5.9). For each point p in $\bar{K}$ there exists a closed cube P containing p such that $P \cap V = \varnothing$. From the diagram

$$
\begin{array}{ccccc}
H_{n-1}(V) & \xrightarrow{\ i_*\ } & H_{n-1}(U) & \xleftarrow[\approx]{\ \Delta\ } & H_n(R^n, U) \\
\downarrow & & \downarrow & & \downarrow{\scriptstyle j_{p*}} \\
H_{n-1}(R^n - p) & \xrightarrow{\ \approx\ } & H_{n-1}(R^n - p) & \xleftarrow[\approx]{\ \Delta\ } & H_n(R^n, R^n - p)
\end{array}
$$

in which the rectangles commute, it is evident that the image of b' under the homomorphism

$$H_{n-1}(V) \to H_{n-1}(R^n - p)$$

is zero. A finite number of such closed cubes cover $\bar{K}$, say $P_1, \ldots, P_m$.

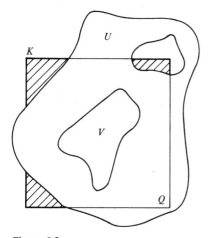

Figure 5.9

Now suppose that the image of b' under the homomorphism

$$H_{n-1}(V) \to H_{n-1}(Q - (P_1 \cup \cdots \cup P_k))$$

is zero (this is certainly true when $k = 0$, since Q is contractible).

Denoting $Q_k = Q - (P_1 \cup \cdots \cup P_k)$, consider the Mayer–Vietoris sequence relating Q_k and $\mathbb{R}^n - P_{k+1}$:

$$\cdots \to H_n(Q_k \cup (\mathbb{R}^n - P_{k+1}))$$

$$\to H_{n-1}(Q_{k+1}) \xrightarrow{\text{mono}} H_{n-1}(Q_k) \oplus H_{n-1}(\mathbb{R}^n - P_{k+1}) \to \cdots.$$

The first group is zero by Lemma 5.1. The images of b' in the two direct summands are both zero; hence, the image of b' in $H_{n-1}(Q_{k+1})$ is zero. This completes the inductive step and we conclude that the image of b' under the homomorphism

$$H_{n-1}(V) \to H_{n-1}(Q - (P_1 \cup \cdots \cup P_m)) \xrightarrow{\text{(incl)}_*} H_{n-1}(Q \cap U)$$

is zero. Since the inclusion of V in U factors through $Q \cap U$, it follows that the image of b' in $H_{n-1}(U)$, that is, b, is zero. Hence, $a = 0$. $\square$

5.4 LEMMA If x and y are points in the interior of a connected n-manifold M, then there is a homeomorphism $h: M \to M$, homotopic to the identity, having $h(x) = y$.

PROOF Note that since M is locally homeomorphic to H^n, M is locally pathwise connected. That is, each point of M is contained in a pathwise

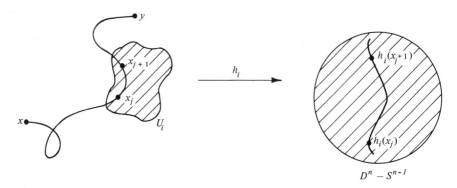

Figure 5.10

connected open set. This implies that the path components of M are both open and closed. Since M is connected, there can only be one path component, so M must be pathwise connected.

Now if M is connected, so is $M - \partial M =$ the interior of M (see the following exercise). So let $p: [0, 1] \to M - \partial M$ be a path from x to y. This compact subset of $M - \partial M$ may be covered by a finite number of open sets, each homeomorphic to the open unit disk in $\mathbb{R}^n$. Denote these disks by $U_1, \ldots, U_m$ and the corresponding homeomorphisms by $h_1, \ldots, h_m$.

Let $x = x_0, x_1, \ldots, x_k = y$ be a collection of points on the path with the property that for each j, the segment from x_j to x_{j+1} is contained in some U_i (Figure 5.10). The desired homeomorphism may now be constructed inductively. It is sufficient then to show that for $0 \leq j < k$ there is a homeomorphism $h: M \to M$, homotopic to the identity, with $h(x_j) = x_{j+1}$.

So suppose x_j and x_{j+1} are in U_i. Define a homeomorphism

$$g: \quad D^n - S^{n-1} \to \mathbb{R}^n$$

by

$$g(z) = \frac{z}{1 - |z|}.$$

The inverse of g is given by

$$g^{-1}(w) = \frac{w}{1 + |w|}.$$

Let $gh_i(x_j) = (a_1, a_2, \ldots, a_n)$ and $gh_i(x_{j+1}) = (b_1, b_2, \ldots, b_n)$. Define the translation function

$$f: \quad \mathbb{R}^n \to \mathbb{R}^n$$

by $f(w_1, \ldots, w_n) = (w_1 + (b_1 - a_1), w_2 + (b_2 - a_2), \ldots, w_n + (b_n - a_n))$. Then f is a homeomorphism with

$$f(gh_i(x_j)) = gh_i(x_{j+1}).$$

Moreover, f is homotopic to the identity via the homotopy

$$f_t(w_1, \ldots, w_n) = (w_1 + t(b_1 - a_1), \ldots, w_n + t(b_n - a_n)), \qquad 0 \leq t \leq 1.$$

Thus, we have a homeomorphism for each t

$$g^{-1} \circ f_t \circ g: \quad D^n - S^{n-1} \to D^n - S^{n-1}$$

that takes $h_i(x_j)$ into $h_i(x_{j+1})$ when $t = 1$. Note further that, for each t, this may be extended to a homeomorphism from D^n to D^n by defining it to be the identity on the boundary (see Exercise 3).

Now define a homotopy $h_t: M \to M$ by

$$h_t(z) = \begin{cases} z & \text{if } z \in M - U_i \\ h_i^{-1} \circ g^{-1} \circ f_t \circ g \circ h_i(z) & \text{if } z \in U_i. \end{cases}$$

Then each h_t is a homeomorphism, h_0 is the identity and $h_1(x_j) = x_{j+1}$. This completes the inductive step, so that by composing maps and homotopies we may give a homeomorphism homotopic to the identity taking x into y. $\square$

NOTE We actually have proved something stronger than the conclusion of the lemma. If $f, g: X \to Y$ are homeomorphisms between topological spaces, then f is *isotopic* to g if there exists a map $F: X \times [0, 1] \to Y$ such that

(1) $F(x, 0) = f(x)$;
(2) $F(x, 1) = g(x)$;
(3) for $0 \leq t \leq 1$ the map $x \to F(x, t)$ is a homeomorphism of X onto Y.

The construction used in the theorem makes it evident that the map h is isotopic to the identity.

Exercise 2 If M is a connected n-manifold, show that the interior of M, $M - \partial M$, is also connected.

Exercise 3 Let $(a_1, \ldots, a_n)$ be a point in R^n and define $f: R^n \to R^n$ by $f(x_1, \ldots, x_n) = (x_1 + a_1, \ldots, x_n + a_n)$. Using the map $g: D^n - S^{n-1} \to R^n$ defined by $g(z) = z/(1 - |z|)$, show that the homeomorphism

$$g^{-1} \circ f \circ g: \quad D^n - S^{n-1} \to D^n - S^{n-1}$$

may be extended to a homeomorphism $D^n \to D^n$ by defining it to be the identity on the boundary.

5.5 THEOREM If M is a connected, noncompact n-manifold without boundary, then $H_n(M) = 0$.

PROOF First note that if p is any point in M, the homomorphism $k_*: H_n(M) \to H_n(M, M - p)$, induced by the inclusion, is identically zero. To check this let $\{z\} \in H_n(M)$ be represented by the cycle z and let $C \subseteq M$ be a compact subset which supports the cycle z.

First suppose $p \in M - C$, so that $\{z\}$ is in the image of the homomorphism $H_n(M - p) \to H_n(M)$. Then by the exact sequence of the pair $(M, M - p)$ it follows that $k_*(\{z\}) = 0$.

If $p \in C$, then select a point q with $q \in M - C$. Such a point must exist, since C is compact and M is not. By Lemma 5.4 there is a homeomorphism $h: M \to M$, homotopic to the identity, with $h(p) = q$. The restriction of h will then yield a homeomorphism from $M - p$ to $M - q$.

From the commutativity of the diagram

$$
\begin{array}{ccc}
H_n(M) & \xrightarrow{\;k_*\;} & H_n(M, M - p) \\
\Big\downarrow{\scriptstyle h_* = \mathrm{id}} & & \Big\downarrow{\scriptstyle \approx} \\
H_n(M) & \xrightarrow{\;k'_*\;} & H_n(M, M - q)
\end{array}
$$

and the fact that $k_*'(\{z\}) = 0$ we conclude that $k_*(\{z\}) = 0$.

Now let $\{z\}$ be an arbitrary homology class as above and cover the compact set C with a finite number of coordinate neighborhoods $U_1, \ldots, U_k$, where each U_i is homeomorphic to an open disk in $\mathbb{R}^n$. Denoting $V_j = \bigcup_{i=1}^{j} U_i$, we will show that $\{z\} = 0$ by proving that $\{z\}$ is a boundary in V_k, that is, $H_n(V_k) = 0$.

The argument proceeds by induction on k. For $k = 1$ the result follows from Lemma 5.1. So suppose $H_n(V_m) = 0$ and consider the Mayer–Vietoris sequence for the union $V_m \cup U_{m+1} = V_{m+1}$:

$$
H_n(V_m) \oplus H_n(U_{m+1}) \to H_n(V_{m+1}) \to H_{n-1}(V_m \cap U_{m+1})
$$
$$
\to H_{n-1}(V_m) \oplus H_{n-1}(U_{m+1}).
$$

Now the first term is zero by the inductive hypothesis and Lemma 5.1 and $H_{n-1}(U_{m+1}) = 0$ since U_{m+1} is homeomorphic to an open disk. Thus, to prove that $H_n(V_{m+1}) = 0$ it is sufficient to show that the homomorphism

$$
i_*: \quad H_{n-1}(V_m \cap U_{m+1}) \to H_{n-1}(V_m)
$$

is a monomorphism.

So suppose that $i_*(\beta) = 0$. Then there exist elements $\beta' \in H_n(U_{m+1}, V_m \cap U_{m+1})$ and $\beta'' \in H_n(V_m, V_m \cap U_{m+1})$ such that $\Delta_1(\beta') = \beta = \Delta_2(\beta'')$, where Δ_1 and Δ_2 are the respective connecting homomorphisms.

Consider the following diagram:

$$H_n(V_m, V_m \cap U_{m+1}) \xrightarrow{\;i_{2*}\;} H_n(V_{m+1}, V_m \cap U_{m+1}) \xrightarrow{\;l_*\;} H_n(M, M - p)$$

with vertical maps Δ_2, i_{1*}, $\approx$, and the diagonal maps j_*, i_{p*} through $H_n(V_{m+1})$:

$$H_{n-1}(V_m \cap U_{m+1}) \xleftarrow{\;\Delta_1\;} H_n(U_{m+1}, V_m \cap U_{m+1}) \xrightarrow{\;j_{p*}\;} H_n(U_{m+1}, U_{m+1} - p)$$

Setting $\bar{\beta} = i_{1*}(\beta') - i_{2*}(\beta'')$, observe that $\Delta(\bar{\beta}) = 0$, where Δ is the connecting homomorphism of the pair $(V_{m+1}, V_m \cap U_{m+1})$. Thus, there exists an $\alpha \in H_n(V_{m+1})$ with $j_*(\alpha) = \bar{\beta}$.

Let $p \in U_{m+1} - V_m \cap U_{m+1}$. Then by the remarks at the beginning of the proof, $i_{p*}(\alpha) = 0$. Thus

$$0 = l_*(\,j_*(\alpha)) = l_*(\bar{\beta}) = l_*(i_{1*}(\beta') - i_{2*}(\beta''))$$
$$= l_* i_{1*}(\beta') - l_* i_{2*}(\beta'').$$

Now since $p \notin V_m$, $l_* i_{2*}$ factors through $H_n(M - p, M - p) = 0$, so $l_* i_{2*}(\beta'') = 0$. Hence, $l_* i_{1*}(\beta') = 0$. This implies that $j_{p*}(\beta') = 0$.

It follows from Lemma 5.3 that β' must be zero, hence $\beta = 0$ and i_* is a monomorphism. Therefore, $H_n(V_{m+1}) = 0$ and we have completed the inductive step. $\square$

5.6 COROLLARY Let M be a closed connected n-manifold. If $z \in H_n(M)$ and $p \in M$ such that

$$i_{p*}: \quad H_n(M) \to H_n(M, M - p)$$

has $i_{p*}(z) = 0$, then $z = 0$.

PROOF Since $i_{p*}(z) = 0$, there must be an element $z' \in H_n(M - p)$ with z being the image of z'. However, $M - p$ is not compact, so by Theorem 5.5 $H_n(M - p) = 0$. Thus, z' and also z must be zero. $\square$

5.7 COROLLARY If M is a connected n-manifold without boundary, then either

(i) $H_n(M) = 0$, or

(ii) $H_n(M) \approx Z$, and for every $p \in M$ the homomorphism $i_{p*}: H_n(M) \to H_n(M, M - p)$ is an isomorphism.

PROOF From Corollary 5.6 we have that i_{p*} is a monomorphism. Since $H_n(M, M - p) \approx Z$, it follows that either $H_n(M) = 0$ or $H_n(M) \approx Z$. So suppose $H_n(M) \neq 0$ and let $z \in H_n(M)$ and $w \in H_n(M, M - p)$ be generators for the respective infinite cyclic groups. Then $i_{p*}(z) = \pm m \cdot w$ for some positive integer m. We must show that $m = 1$.

Note that the same proof as for Theorem 5.5 may be given to show that for any abelian group G, $H_n(M; G) = 0$ for M a connected noncompact manifold without boundary. Then consider the diagram

$$
\begin{array}{ccc}
H_n(M) \otimes Z_m & \xrightarrow{\ i_{p*} \otimes \mathrm{id}\ } & H_n(M, M - p) \otimes Z_m \\
{\scriptstyle \alpha_1} \downarrow {\scriptstyle \mathrm{mono}} & & {\scriptstyle \alpha_2} \downarrow {\scriptstyle \mathrm{mono}} \\
H_n(M; Z_m) & \xrightarrow[\ \mathrm{mono}\ ]{\ i_{p*}\ } & H_n(M, M - p; Z_m)
\end{array}
$$

where the vertical monomorphisms come from the universal coefficient theorem. The commutativity of the square implies that $i_{p*} \otimes \mathrm{id}$ is a monomorphism. But

$$(i_{p*} \otimes \mathrm{id})(z \otimes 1) = i_{p*}(z) \otimes 1 = \pm m \cdot w \otimes 1 = \pm w \otimes m = 0,$$

so that $z \otimes 1 = 0$. This only happens if $m = 1$. Therefore, i_{p*} is an isomorphism. $\square$

Let M be an n-manifold without boundary. For each $p \in M$ let

$$T_p = H_n(M, M - p) \approx Z$$

and for coefficients in Z_2

$$T_p(Z_2) = H_n(M, M - p; Z_2) \approx Z_2.$$

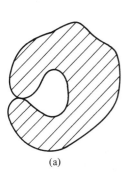
(a)

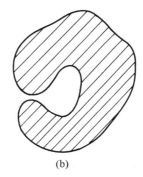
(b)

Figure 5.11

Define the set

$$\mathfrak{I} = \bigcup_{p \in M} T_p.$$

We want to introduce a topology on the set $\mathfrak{I}$. To do so requires the notion of a proper n-ball. A *proper n-ball* in M is an open set $V \subseteq M$ such that there exists a homeomorphism of D^n onto $\bar{V}$ taking S^{n-1} onto $\bar{V} - V$. For example, the interior of the region shown in Figure 5.11a fails to be a proper n-ball while the interior of the region in Figure 5.11b is a proper n-ball.

Exercise 4 Show that if M is an n-manifold without boundary, then the collection of all proper n-balls in M forms a basis for the topology on M.

Now if V is a proper n-ball in M and $p \in V$, then there is an isomorphism

$$j_{p*}: \quad H_n(M, M - V) \xrightarrow{\approx} T_p.$$

As a basis for the topology on $\mathfrak{I}$ we take the sets

$$U_{\alpha, V} = \{ j_{p*}(\alpha) \mid p \in V \}$$

as V ranges over all proper n-balls of M and α ranges over all elements for $H_n(M, M - V)$. For example, the choice of a generator for $H_n(M, M - V)$ dictates, via the isomorphisms j_{p*}, a generator for each T_p, and these selected generators form a sheet in $\mathfrak{I}$ which is homeomorphic to V. Since this may be done for either generator of $H_n(M, M - V)$ we see that the generators of the T_p form two disjoint sheets each homeomorphic to V.

More generally, if $\tau: \mathfrak{I} \to M$ is the natural projection, it is evident that τ is a local homeomorphism. Each component of $\mathfrak{I}$ is a covering space of M with either one or two sheets. In particular the generators of all the T_p form either a double covering of M or two distinct simple coverings. The restriction of τ to this subset of $\mathfrak{I}$ is the *orientation double covering* of M. In the case that there are two distinct simple coverings we say that M is *orientable*. An *orientation* of M is a map $s: M \to \mathfrak{I}$ with $\tau \circ s = $ identity on M and $s(p)$ a generator of T_p for each $p \in M$. Of course, an orientable manifold has two possible orientations.

Exercise 5 Let B be the Möbius band so that $M = B - \partial B$ is a 2-manifold without boundary. Prove that M is not orientable by showing that the domain of its orientation double covering is homeomorphic to the annulus $S^1 \times (0, 1)$.

Exercise 6   Let M be an n-manifold without boundary and let $\hat{M}$ be the domain of its orientation double covering. Then show that $\hat{M}$ is an orientable n-manifold.

Following the same procedure for the groups $T_p(Z_2)$, the generators form a simple covering of M so that there always exists a unique Z_2-orientation.

If M is a closed n-manifold, a *fundamental class* on M is an element $z \in H_n(M)$ such that

$$i_{p*}: \quad H_n(M) \to H_n(M, M - p) = T_p$$

has $i_{p*}(z)$ a generator of T_p for each $p \in M$. A cycle representing z is a *fundamental cycle*.

5.8   LEMMA   Let M be a closed n-manifold and U be an open subset of M. If an element $x \in H_n(M, U)$ has $j_{p*}(x) = 0$ for all $p \in M - U$, where

$$j_{p*}: \quad H_n(M, U) \to T_p,$$

then $x = 0$.

PROOF   First suppose that $M - U$ is contained in some coordinate neighborhood W. Then consider the commutative diagram

$$\begin{array}{ccc} H_n(W, W \cap U) & \xrightarrow{\ h_{p*}\ } & H_n(W, W - p) \\ \approx \Big\downarrow & & \approx \Big\downarrow \\ H_n(M, U) & \xrightarrow{\ j_{p*}\ } & H_n(M, M - p) = T_p \end{array}$$

where the vertical homomorphisms are excision isomorphisms. For each $p \in M - U = W - (W \cap U)$, it follows that $j_{p*}(x) = 0$ if and only if h_{p*} kills the preimage of x. Then by Lemma 5.3 the preimage of x must be zero, which implies that $x = 0$.

For the general case, since $M - U$ is compact, we express $M - U$ as the union of a finite number of compact sets, each contained in a coordinate neighborhood. We proceed by induction on the number k of such compact sets. In the previous paragraph we have proved the result for $k = 1$. For the inductive step we use the Mayer–Vietoris sequence

$$H_{n+1}(M, U' \cup U'') \to H_n(M, U' \cap U'') \to H_n(M, U') \oplus H_n(M, U''),$$

where U' and U'' are open sets in M, and the fact that $H_{n+1}(M, U' \cup U'')=0$, which follows by Lemmas 5.1 and 5.5 and the exact sequence of a pair. $\square$

5.9   LEMMA   Let M be a closed orientable n-manifold with orientation $s: M \to \mathfrak{I}$. Then there exists a class $z \in H_n(M)$ such that $i_{p*}(z) = s(p) \in T_p$ for all $p \in M$.

PROOF   From our previous observations we know that for each p there exists a proper n-ball V_p about p and an element $x_p \in H_n(M, M - \bar{V}_p)$ such that if $q \in \bar{V}_p$, $j_{q*}(x_p) = s(q)$. The technique then is to piece together such proper n-balls, using a Mayer–Vietoris sequence and the compactness of M, to construct the desired global homology class.

Since M is compact, there is a finite collection $V_1, \ldots, V_k$ of proper n-balls which cover M. Suppose that there is an element

$$z_m \in H_n(M, M - (\bar{V}_1 \cup \cdots \cup \bar{V}_m))$$

such that

$$j_{q*}(z_m) = s(q)$$

for all $q \in \bar{V}_1 \cup \cdots \cup \bar{V}_m$. Then consider the relative Mayer–Vietoris sequence

$$H_n(M, M - (\bar{V}_1 \cup \cdots \cup \bar{V}_{m+1}))$$
$$\to H_n(M, M - (\bar{V}_1 \cup \cdots \cup \bar{V}_m)) \oplus H_n(M, M - \bar{V}_{m+1})$$
$$\to H_n(M, M - (\bar{V}_1 \cup \cdots \cup \bar{V}_m) \cap \bar{V}_{m+1}).$$

Starting with the element $z_m - x_{m+1}$ in the direct sum, let w be its image in $H_n(M, M - (\bar{V}_1 \cup \cdots \cup \bar{V}_m) \cap \bar{V}_{m+1})$. This implies that for all $q \in (\bar{V}_1 \cup \cdots \cup \bar{V}_m) \cap \bar{V}_{m+1}$, $j_{q*}(w) = 0$. Now by Lemma 5.8 it follows that $w = 0$.

Let $z_{m+1} \in H_n(M, M - (\bar{V}_1 \cup \cdots \cup \bar{V}_{m+1}))$ be the element which is mapped into $z_m - x_{m+1}$ and note that $j_{q*}(z_{m+1}) = s(q)$ for all $q \in \bar{V}_1 \cup \cdots \cup \bar{V}_{m+1}$. This completes the inductive step and the desired class z is z_k. $\square$

All of these results are summarized in the following theorem expressing the precise relation between orientation and fundamental class:

5.10   THEOREM   If M is a closed connected orientable n-manifold with orientation $s: M \to \mathfrak{I}$, then there is a unique fundamental class $z \in H_n(M)$ such that $i_{p*}(z) = s(p)$ for each $p \in M$.

PROOF This follows immediately from the previous lemmas. From Lemma 5.9 we have the existence of such a fundamental class z. If z' is another such class, then for each $p \in M$

$$i_{p*}(z - z') = s(p) - s(p) = 0$$

so that $z - z' = 0$ by Corollary 5.6. This proves uniqueness. $\square$

Exercise 7 Let M_1 be a closed orientable n_1-manifold and M_2 be a closed orientable n_2-manifold. Then show $M_1 \times M_2$ is a closed orientable $(n_1 + n_2)$-manifold.

Our procedure for proving the Poincaré duality theorem follows the style of Milnor [1957] by first establishing a form of the Thom isomorphism. So for the present, we assume that M is a closed, orientable n-manifold with an orientation $s: M \to \mathcal{S}$. Then by the above exercise, $M \times M$ is a compact, orientable $2n$-manifold. Define maps $\pi_1, \pi_2: M \times M \to M$ by projection onto the first or second coordinate, and for any $p \in M$

$$l_p, r_p: \quad M \to M \times M$$

by $l_p(x) = (p, x)$, $r_p(x) = (x, p)$. Finally, denote by

$$\Delta: \quad M \to M \times M$$

the diagonal $\Delta(x) = (x, x)$ and note that

$$\pi_1 \circ \Delta = \pi_2 \circ \Delta = \text{identity on } M.$$

5.11 LEMMA Let V be a proper n-ball in M with $p \in V$ corresponding to the origin in D^n. There is a homeomorphism

$$\theta: \pi_1^{-1}(V) = V \times M \to \pi_1^{-1}(V)$$

such that

(i) $\pi_1 = \pi_1 \circ \theta$ on all of $\pi_1^{-1}(V)$;

(ii) $\theta \circ \Delta = r_p$ on V;

(iii) $(\pi_2 \circ \theta \circ l_q)_*(s(q)) = s(p)$ for all $q \in V$.

NOTE This states that $V \times M$ may be deformed in such a way that the first coordinate is unchanged (i), and the diagonal over V is transformed into the level set $V \times \{p\}$ (ii). Furthermore this is done in such a way as to preserve the orientation in the sense that the composition

$$H_n(M, M - q) \xrightarrow{l_{q*}} H_n(q \times M, q \times (M - q))$$
$$\xrightarrow{\theta_*} H_n(q \times M, q \times (M - p))$$
$$\xrightarrow{\pi_{2*}} H_n(M, M - p)$$

takes $s(q)$ into $s(p)$ for all q in V.

PROOF Denote by

$$h: \quad \bar{V} \to D^n$$

the homeomorphism taking p into the origin. Define a homeomorphism

$$\lambda: \quad (D^n - \partial D^n) \times D^n \to (D^n - \partial D^n) \times D^n$$

as follows: for $x \in D^n - \partial D^n$, $y \in \partial D^n$, map the entire segment from (x, x) to (x, y) linearly into the segment from $(x, 0)$ to (x, y) (Figure 5.12).
   Now define

$$\theta(q, q') = \begin{cases} (q, q') & \text{if} \quad q' \notin V \\ (q, h^{-1}(\lambda(h(q), h(q')))) & \text{for} \quad q' \in V. \end{cases}$$

For fixed $q \in V$, as $q' \in V$ approaches ∂V, $\theta(q, q')$ approaches (q, q'). It follows that θ is a well-defined homeomorphism with the desired properties. $\square$

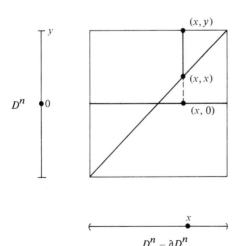

Figure 5.12

For any open set $U \subseteq M$ denote by $U^\times$ the pair

$$U^\times = (\pi_1^{-1}(U), \pi_1^{-1}(U) - \Delta(M)) = (U \times M, U \times M - \Delta(M))$$

and in particular

$$M^\times = (M \times M, M \times M - \Delta(M)).$$

5.12   LEMMA

(i) $H_i(M^\times) = 0$ for $i < n$;

(ii) $H_0(M) \approx H_n(M^\times)$ under the homomorphism sending the 0-chain represented by $p \in M$ into the relative class represented by $l_{p*}(s(p))$, where

$$l_{p*}: \quad H_n(M, M - p) \to H_n(M \times M, M \times M - \Delta(M)).$$

PROOF   First consider the statement of the lemma with a proper n-ball V replacing M. The homeomorphism θ of Lemma 5.11 induces

$$\theta: \quad V^\times = (V \times M, V \times M - \Delta(M)) \to V \times (M, M - p),$$

since $\Delta(V)$ is taken into $V \times p$. Therefore, the induced homomorphism on the relative homology groups is an isomorphism. Applying the Künneth formula of Theorem 4.5 to $V \times (M, M - p)$, Statement (i) follows for $V^\times$.

Consider the composition

$$H_n(V, V - q) \xrightarrow{\ l_{q*}\ } H_n(V \times M, V \times M - \Delta(M))$$

$$\xrightarrow[\approx]{\ \theta_*\ } H_n(V \times (M, M - p)) \xrightarrow{\ \pi_{2*}\ } H_n(M, M - p).$$

$$\wr$$

$$H_0(V) \otimes H_n(M, M - p)$$

From Lemma 5.11, Part (iii) we know that $\pi_{2*}\theta_*l_{q*}(s(q)) = s(p)$. The vertical isomorphism follows from the Künneth formula, where $H_0(V) \otimes H_n(M, M - p)$ is the infinite cyclic group generated by $\{q\} \otimes s(p)$. These two isomorphisms imply that Part (ii) holds for V replacing M.

Finally, we again use an inductive procedure to extend to the general case. Suppose that U and V are open sets in M such that the lemma holds for U, V, and $U \cap V$. There is a diagram of Mayer–Vietoris sequences

$$\cdots \to H_0(U \cap V) \to H_0(U) \oplus H_0(V) \to H_0(U \cup V) \to 0 \to \cdots$$

$$\downarrow \qquad\qquad \downarrow \qquad\qquad \downarrow \qquad\qquad \downarrow$$

$$\cdots \to H_n((U \cap V)^\times) \to H_n(U^\times) \oplus H_n(V^\times) \to H_n((U \cup V)^\times) \to H_{n-1}((U \cap V)^\times) \to \cdots,$$

where the vertical homomorphisms are those described in Part (ii). Note that if p and q are in the same path component of U, then it follows from Lemma 5.11, Part (iii) that $l_{p*}(s(p)) = l_{q*}(s(q))$. Applying the five lemma (Exercise 4, Chapter 2) completes the inductive step and, since M is compact, the proof is complete. $\square$

We are now ready to prove the following important theorem which has many applications in algebraic topology.

5.13   THEOREM (*Thom isomorphism theorem*)   For a compact oriented n-manifold M without boundary, there is a cohomology class $U \in H^n(M^\times)$ such that for any coefficient group G the homomorphism

$$\Phi^*:\quad H^k(M; G) \to H^{n+k}(M^\times; G)$$

given by

$$\Phi^*(x) = U \cup \pi_1^*(x)$$

is an isomorphism. Here the cup product has the form

$$H^k(M \times M; G) \otimes H^n(M \times M, M \times M - \Delta(M); Z)$$
$$\to H^{n+k}(M \times M, M \times M - \Delta(M); G).$$

NOTE   The class $U \in H^n(M^\times)$ is called the *Thom class* of the topological manifold M.

PROOF   We prove the theorem for $G = Z$. The general case follows by applying the universal coefficient theorem.

Since $H_i(M^\times) = 0$ for $i < n$, it follows from the universal coefficient theorem that $H^i(M^\times) = 0$ for $i < n$ and $H^n(M^\times) \approx \text{Hom}(H_n(M^\times), Z)$. From Lemma 5.12 we have a natural isomorphism of $H_n(M^\times)$ with $H_0(M)$, hence also of $\text{Hom}(H_n(M^\times), Z)$ with $\text{Hom}(H_0(M), Z)$. Then we define $U \in H^n(M^\times)$ to be the class corresponding to the augmentation homomorphism

$$H_0(M) \to Z$$

under these isomorphisms. In particular then it follows from Lemma 5.12 that for all $p \in M$ the Kronecker index

$$\langle U, l_{p*}(s(p)) \rangle = 1.$$

For any open set $V \subseteq M$ we denote by $U_v \in H^n(V^\times)$ the restriction of the Thom class U. There is a cap product

$$H^n(V \times M, V \times M - \Delta(M)) \otimes H_k(V \times M, V \times M - \Delta(M)) \to H_{k-n}(V \times M)$$

sending $U_v \otimes \alpha$ into $U_v \cap \alpha$. For any $\alpha \in H_k(V^\times)$ define $\Phi_*(\alpha)$ to be the element of $H_{k-n}(V)$ given by $\Phi_*(\alpha) = \pi_{1*}(U_v \cap \alpha)$. Thus

$$\Phi_*: \quad H_*(V^\times) \to H_*(V)$$

is a natural homomorphism of degree $-n$ between graded groups.

Now restrict to the case of V being a proper n-ball. Under the isomorphism of Lemma 5.11

$$\theta^*: \quad H^n(V \times (M, M - p)) \to H^n(V^\times),$$

the element $\theta^{*-1}(U_v)$ may be identified via the Künneth formula with $1 \otimes \gamma \in H^0(V) \otimes H^n(M, M - p)$, where γ is a generator of the infinite cyclic group $H^n(M, M - p)$.

By the naturality of the cap product

$$\pi_{1*}(U_v \cap \alpha) = \pi_{1*} \circ \theta_*(\theta^*(\theta^{*-1}(U_v)) \cap \alpha)$$

$$= \pi_{1*}(\theta^{*-1}(U_v) \cap \theta_*(\alpha)).$$

So if $\omega \in H_n(M, M - p)$ is a generator with $\langle \gamma, \omega \rangle = 1$, then

$$\pi_{1*}(U_v \cap \alpha) = \beta,$$

where $\beta \otimes \omega$ corresponds to α under the isomorphisms

$$H_{k-n}(V) \otimes H_n(M, M - p) \xrightarrow{\approx} H_k(V \times (M, M - p)) \xleftarrow[\approx]{\theta_*} H_k(V^\times).$$

Therefore, $\Phi_*: H_k(V^\times) \to H_{k-n}(V)$ is an isomorphism for each k.

If $V_1 \subseteq V_2$ are open sets in M, the restriction of U_{v_2} is U_{v_1}. This fact, together with naturality and the Mayer–Vietoris sequence, may be used in the manner of Lemma 5.12 to extend the result inductively to an isomorphism

$$\Phi_*: \quad H_k(M^\times) \xrightarrow{\approx} H_{k-n}(M).$$

By returning to the chain level we can define the adjoint of Φ_*; this yields a homomorphism

$$\Phi^*: \quad H^i(M) \to H^{i+n}(M^\times).$$

Applying the universal coefficient theorem, we see that Φ^* is also an iso-morphism. Finally note that

$$\langle \Phi^*(x), y \rangle = \langle x, \Phi_*(y) \rangle$$
$$= \langle x, \pi_{1*}(U \cap y) \rangle$$
$$= \langle \pi_1^* x, U \cap y \rangle$$
$$= \langle U \cup \pi_1^*(x), y \rangle$$

for any x and y, so that $\Phi^*(x) = U \cup \pi_1^*(x)$. □

Example As an aid to understanding this important theorem, consider the following simple example: Let $M = S^1$ so that $M \times M$ is the two-dimensional torus. Recall from Chapter 4 the determination of the cohomology ring structure in $M \times M$. As before we denote generating 1-cycles in $M \times M$ by $\bar\alpha$ and $\bar\beta$, and their dual cocyles by α and β (Figure 5.13). It is not difficult to see that $M \times M - \Delta M$ has the homotopy type of S^1, as is demonstrated in the following deformation (Figure 5.14).

Now given an orientation for $M = S^1$, there is a generator $\bar{a} \in H_1(M)$ such that $i_{p*}(\bar{a}) = s(p)$ for all $p \in M$. Thus, from the commutative

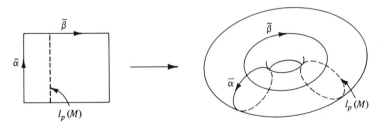

Figure 5.13

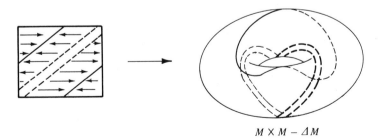

$M \times M - \Delta M$

Figure 5.14

diagram

$$\begin{array}{ccc} H_1(M) & \xrightarrow{\;i_{p*}\;} & H_1(M, M-p) \\ \downarrow{\scriptstyle l_{p*}} & & \downarrow{\scriptstyle l_{p*}} \\ H_1(M \times M) & \xrightarrow{\;i_*\;} & H_1(M^\times) \end{array}$$

we see that $l_{p*}(s(p)) = l_{p*}i_{p*}(\bar{a}) = i_* l_{p*}(\bar{a})$. With a possible change in sign resulting from the choice of orientation, we have $l_{p*}(\bar{a}) = \bar{a}$. Thus, the Thom class $U \in H^1(M^\times)$ will have the property that

$$1 = \langle U, l_{p*}s(p) \rangle = \langle U, i_*(\bar{a}) \rangle = \langle i^*U, \bar{a} \rangle.$$

From the exact cohomology sequence of the pair $M^\times$,

$$H^1(M^\times) \xrightarrow{\;i^*\;} H^1(M \times M) \xrightarrow{\;j^*\;} H^1(M \times M - \Delta M),$$

it is not difficult to argue that i^* is a monomorphism and j^* is an epimorphism. Furthermore, $H^1(M \times M)$ is free abelian with basis elements α and β, and the kernel of j^* is infinite cyclic generated by $\alpha - \beta$. This uniquely determines the Thom class U corresponding to the given orientation. Changing the orientation of M changes the sign of U.

Finally consider the Thom homomorphism

$$\Phi^*: \quad H^1(M) \to H^2(M^\times).$$

If a is the generator dual to $\bar{a}$, $\Phi^*(a) = U \cup \pi_1^*(a)$. Now in $H^1(M \times M)$ we have

$$\langle \pi_1^*(a), m\bar{\alpha} + n\bar{\beta} \rangle = \langle a, \pi_{1*}(m\bar{\alpha} + n\bar{\beta}) \rangle = m,$$

so that $\pi_1^*(a) = \alpha$. Using the isomorphism $H^2(M^\times) \xrightarrow{\;i^*\;} H^2(M \times M)$, it follows that

$$i^*(\Phi^*(a)) = i^*(U \cup \pi_1^*(a)) = i^*(U \cup \alpha) = i^*(U) \cup \alpha$$
$$= (\alpha - \beta) \cup \alpha = -\beta \cup \alpha,$$

which is a generator of $H^2(M \times M)$. Thus, $\Phi^*(a)$ is a generator of $H^2(M^\times)$ and Φ^* is an isomorphism.

At this point we need another very basic property of topological manifolds.

5.14 THEOREM If M is a closed topological n-manifold, there exists a topological embedding of M in $\mathbb{R}^k$ for some large value of k. Furthermore, under this embedding there exists an open set U about M such that M is a retract of U, that is, there exists a map $r: U \to M$ such that $r|_M$ is the identity.

PROOF See Appendix II. ☐

5.15 LEMMA For M a closed manifold, there exists a neighborhood N of $\Delta(M)$ in $M \times M$ such that $\pi_1|_N$ and $\pi_2|_N$ are homotopic as maps from N to M.

PROOF Applying Theorem 5.14 we embed M in $\mathbb{R}^k$ and let U be a neighborhood of M which retracts onto M. Since M is compact there exists an $\varepsilon > 0$ such that for any points x, $y \in M$ having distance (in $\mathbb{R}^k$) between x and y less than ε, the segment from x to y lies in U. It is evident then that any two maps into M with the distance between corresponding points less than ε are homotopic in U via the obvious homotopy. Applying the retraction r moves the homotopy into M.

Now the projection maps π_1 and π_2 coincide on $\Delta(M)$ in $M \times M$. Again using compactness, there must exist a neighborhood N of $\Delta(M)$ in $M \times M$ such that the distance between $\pi_1|_N$ and $\pi_2|_N$ is less than ε. It follows then that these restrictions are homotopic in M. ☐

5.16 LEMMA Define $t: M \times M \to M \times M$ by $t(x, y) = (y, x)$ and note that t induces a map of pairs $t: M^\times \to M^\times$. Then for $x \in H^*(M^\times; G)$, $t^*(x) = (-1)^n x$.

PROOF First let V be a proper n-ball in M and consider the diagram

$$
\begin{array}{ccc}
H^n(M \times M, M \times M - \Delta(M)) & \xrightarrow{\;t^*\;} & H^n(M \times M, M \times M - \Delta(M)) \\
\downarrow{\scriptstyle i^*} & & \downarrow{\scriptstyle i^*} \\
H^n(V \times V, V \times V - \Delta(V)) & \xrightarrow{\;t^*\;} & H^n(V \times V, V \times V - \Delta(V)),
\end{array}
$$

where the vertical homomorphisms are induced by the inclusion map. Note that $H^n(V \times V, V \times V - \Delta(V))$ is infinite cyclic since $(V \times V, V \times V - \Delta(V))$ has the homotopy type of (D^n, S^{n-1}). Furthermore, $i^*(U)$ is a generator of this group and $t^*(i^*U) = (-1)^n i^*(U)$. Thus, we conclude that

$$t^*(U) = (-1)^n U.$$

Now let N be a closed neighborhood of $\Delta(M)$ satisfying the requirement of Lemma 5.15. We may further require that N be invariant under t. Then the diagram

$$H^*(M) \xrightarrow{\pi_1^*} H^*(M \times M) \xrightarrow{U \cup} H^*(M^\times)$$

with $(\pi_1|_N)^*$, j^*, and $j^* \approx$ excision

$$H^*(N) \xrightarrow{j^*(U)\cup} H^*(N, N - \Delta(M))$$

is commutative where the composition across the top is Φ^*. Recalling that $\pi_2 = \pi_1 \circ t$ we have

$$t^* j^* (\Phi^*(x)) = t^* j^* (U \cup \pi_1^*(x)) = t^* (j^*(U) \cup (\pi_1|_N)^*(x))$$

$$= t^* j^*(U) \cup (\pi_2|_N)^*(x) = (-1)^n j^*(U) \cup (\pi_2|_N)^*(x)$$

which by Lemma 5.15

$$= (-1)^n j^*(U) \cup (\pi_1|_N)^*(x) = (-1)^n j^*(\Phi^*(x)).$$

Since j^* and Φ^* are isomorphisms, the result follows. $\square$

5.17 PROPOSITION   If M is a closed n-manifold, then $H_*(M)$ is finitely generated.

PROOF   Using Theorem 5.14 we embed M in a high-dimensional euclidean space $\mathbb{R}^m$ so that some open set N about M in $\mathbb{R}^m$ admits a retraction onto M, $r: N \to M$. Choose a large m-simplex s^m in $\mathbb{R}^m$ so that M is contained in its interior. By the results of Appendix I there exists an integer k so that mesh $\mathrm{Sd}^k s^m$ is less than the distance from M to $\mathbb{R}^m - N$.

Let K be the union of all simplices in $\mathrm{Sd}^k s^m$ whose closures intersect M. Then K is a finite CW complex, $M \subseteq K \subseteq N$ and the retraction r restricts to a retraction $r: K \to M$.

By Proposition 2.23 $H_*(K)$ is finitely generated and by Corollary 1.12 $H_*(M)$ is isomorphic to a direct summand of $H_*(K)$; hence, $H_*(M)$ is finitely generated. $\square$

We are now ready to prove the main theorem of this chapter.

5.18 POINCARÉ DUALITY THEOREM   For M a compact connected orientable n-manifold without boundary, with orientation $s: M \to \mathfrak{I}$ and associated fundamental class z, the homomorphism

$$D: \quad H^k(M; G) \to H_{n-k}(M; G),$$

given by $D(x) = x \cap z$, is an isomorphism for each k.

PROOF   Let $\tilde{z}$ be a cycle in $S_n(M)$ representing z. Then there is a homomorphism of chain complexes

$$D_{\#}: \quad S^k(M; G) \to S_{n-k}(M; G)$$

given by cap product with $\tilde{z}$. Note that $D_{\#}$ commutes with coboundary and boundary operators up to sign and induces D on cohomology and homology.

Now let R be a ring with unit, $x, y \in H^*(M; R)$ and $\alpha, \beta \in H_*(M; R)$. Then there are the elements

$$\alpha \times \beta \in H_*(M \times M; R) \quad \text{and} \quad x \times y = \pi_1^*(x) \cup \pi_2^*(y) \in H^*(M \times M; R).$$

If $U \in H^n(M^\times)$ is the Thom class of M, denote by $\tilde{U}$ the class $i^*(U)$, where

$$i^*: \quad H^n(M^\times) \to H^n(M \times M)$$

is induced by inclusion. Let ε be a chosen generator for $H_0(M)$.

The homomorphism $i_*: H_n(M \times M) \to H_n(M^\times)$ takes $\varepsilon \times z$ into $l_{p*}(s(p))$. Thus

$$(5.19) \qquad \begin{aligned} (-1)^n \langle \tilde{U}, z \times \varepsilon \rangle &= \langle \tilde{U}, \varepsilon \times z \rangle \\ &= \langle i^*(U), \varepsilon \times z \rangle \\ &= \langle U, l_{p*}(s(p)) \rangle \\ &= 1. \end{aligned}$$

If $x \in H^r(M; R)$ and $y \in H^s(M; R)$, then we claim that

$$(5.20) \qquad \tilde{U} \cup (x \times y) = (-1)^{rs} \tilde{U} \cup (y \times x).$$

In order to give this meaning we must view $\tilde{U}$ as an element of $H^n(M \times M; R)$. This is done by using the coefficient homomorphism $Z \to R$ given by taking 1 into the unit of the ring R.

Then in the diagram

$$H^n(M^\times; R) \otimes H^{r+s}(M \times M; R) \xrightarrow{\ \cup\ } H^{n+r+s}(M^\times; R)$$

$$\downarrow i_* \otimes \mathrm{id} \qquad\qquad\qquad\qquad\qquad \downarrow i_*$$

$$H^n(M \times M; R) \otimes H^{r+s}(M \times M; R) \xrightarrow{\ \cup\ } H^{n+r+s}(M \times M; R),$$

we have

$$(-1)^n U \cup (x \times y) = t^*(U \cup (x \times y))$$
$$= t^*(U) \cup t^*(x \times y)$$
$$= (-1)^{n+rs} U \cup (y \times x).$$

Therefore

$$\tilde{U} \cup (x \times y) = i^*(U) \cup \mathrm{id}(x \times y) = i^*(U \cup (x \times y))$$
$$= (-1)^{rs} i^*(U \cup (y \times x)) = (-1)^{rs} \tilde{U} \cup (y \times x)$$

which proves Equation (5.20).

Finally, for $x \in H^k(M; R)$ and $\alpha \in H_k(M; R)$ we have

$$\langle \tilde{U}, D(x) \times \alpha \rangle = \langle \tilde{U}, (x \cap z) \times \alpha \rangle$$
$$= \langle \tilde{U}, (x \times 1) \cap (z \times \alpha) \rangle$$
$$= \langle (x \times 1) \cup \tilde{U}, (z \times \alpha) \rangle.$$

Then by Equation (5.20)

$$\langle \tilde{U}, D(x) \times \alpha \rangle = \langle (1 \times x) \cup \tilde{U}, z \times \alpha \rangle$$
$$= \langle \tilde{U}, (1 \times x) \cap (z \times \alpha) \rangle$$
$$= (-1)^{nk} \langle \tilde{U}, z \times \langle x, \alpha \rangle \cdot \varepsilon \rangle.$$

It follows from Equation (5.19) that this last expression is equal to $(-1)^{n(n-k)} \langle x, \alpha \rangle \cdot 1$. We summarize these statements in the following important equation:

$$(5.21) \qquad\qquad \langle x, \alpha \rangle = (-1)^{n(n-k)} \langle \tilde{U}, D(x) \times \alpha \rangle.$$

For the special case $R = Z_p$, where p is a prime, the universal coefficient theorem becomes an isomorphism,

$$H^*(M; Z_p) \approx \mathrm{Hom}_{Z_p}(H_*(M; Z_p), Z_p).$$

Applying this to the above equation we see that $x \neq 0$ implies $D(x) \neq 0$. Therefore

$$D: \quad H^*(M; Z_p) \to H_*(M; Z_p)$$

is a monomorphism. By Proposition 5.17 these are finite-dimensional vector spaces, so since their dimensions must be the same, D must be an isomorphism.

To extend this result to more general coefficient groups we use the method of "algebraic mapping cylinders" [Eilenberg and Steenrod, 1952]. Recall that $D_\#$ is a homomorphism

$$D_\#: \quad S^k(M) \to S_{n-k}(M)$$

with $D_\# \circ \delta = \partial \circ D_\#$. Define

$$C_{n-k} = S^{k+1}(M) \oplus S_{n-k}(M) \quad \text{or} \quad C_m = S^{n-m+1}(M) \oplus S_m(M),$$

and a boundary operator

$$\bar{\partial}: \quad C_m \to C_{m-1} = S^{n-m+2}(M) \oplus S_{m-1}(M),$$

by

$$\bar{\partial}(\alpha, \beta) = (-\delta\alpha, \partial\beta + D_\#(\alpha)).$$

Then we check that

$$\begin{aligned}
\bar{\partial} \circ \bar{\partial}(\alpha, \beta) &= \bar{\partial}(-\delta\alpha, \partial\beta + D_\#(\alpha)) \\
&= (\delta\delta\alpha, \partial(\partial\beta + D_\#(\alpha)) - D_\#\delta(\alpha)) \\
&= (0, \partial D_\#(\alpha) - D_\#\delta(\alpha)) \\
&= 0.
\end{aligned}$$

Hence, $\{C_m, \bar{\partial}\}$ defines a chain complex C.

There is a short exact sequence

$$0 \to S_m(M) \xrightarrow{\text{incl}} C_m \xrightarrow{\text{proj}} S^{n-m+1}(M) \to 0$$

that determines a short exact sequence of chain complexes (the second homomorphism will be a chain map up to sign only, but this will be sufficient for our purposes). Therefore, we have a long exact sequence of homology groups. What is the connecting homomorphism?

Let $y \in S^{n-m+1}(M)$ with $\delta y = 0$. Pick the element $(y, 0) \in C_m$ that projects onto y. Then

$$\bar{\partial}(y, 0) = (-\delta y, D_\#(y)) = (0, D_\#(y))$$

and the element in $S_{m-1}(M)$ having this as its image is $D_\#(Y)$. Thus the connecting homomorphism for the sequence is D and the sequence has the form

$$\cdots \to H^{n-m}(M) \xrightarrow{D} H_m(M) \to H_m(C) \to H^{n-m+1}(M) \xrightarrow{D} H_{m-1}(M) \to \cdots.$$

Now we may identify $S^k(M; Z_p)$ with $S^k(M) \otimes Z_p$, so that for each prime p we have a long exact sequence

$$\cdots \to H^{n-m}(M; Z_p) \xrightarrow[\approx]{D} H_m(M; Z_p) \to H_m(C; Z_p) \to H^{n-m+1}(M; Z_p) \to \cdots.$$

In the sequence, D is an isomorphism wherever it occurs. Hencs, $H_m(C; Z_p) = 0$ for all integers m and primes p. But since $H_m(C)$ it finitely generated, it follows from the universal coefficient theorem thae $H_m(C) = 0$ for all m. Thus, the first sequence shows D to be an isomorphism for integral coefficients. This same technique shows D to be an isomorphism for G any finitely generated abelian group.

Finally, to extend to the general case, the fact that $H_*(M)$ is finitely generated implies that $H_*(M; G)$ is the direct limit of $\{H_*(M; G')\}$ where G' ranges over the finitely generated subgroups of G. Since D commutes with coefficient homomorphisms, we conclude that D is an isomorphism for general abelian groups G. $\square$

NOTE The essence of the proof is the relationship between the Thom class and the duality homomorphism. Specifically, if x is a cohomology class and α is an homology class, the Kronecker index of x on α is, up to sign, the same as the Kronecker index of the "restricted" Thom class $\tilde{U}$ on the external homology product $D(x) \times \alpha$.

Since all manifolds are orientable when the coefficient group is Z_2, we may duplicate the previous proof to establish:

5.22 THEOREM For M a closed connected n-manifold with Z_2-fundamental class $z_2 \in H_n(M; Z_2)$, the homomorphism

$$D: \quad H^k(M; Z_2) \to H_{n-k}(M; Z_2),$$

given by $D(x) = x \cap z_2$, is an isomorphism. $\square$

We now turn to the relative case, that is, the duality theorem for a manifold with boundary. Therefore, let M be a compact manifold with boundary ∂M. The structures that were defined previously, the local homology groups T_p and the orientation covering $\mathfrak{I}$ with projection τ, may still be defined for points $p \in M - \partial M$. We define $(M, \partial M)$ to be *orientable* if there exists a continuous map $s: M - \partial M \to \mathfrak{I}$ with $\tau \circ s = $ identity and $s(p)$ a generator of T_p for each p in $M - \partial M$.

One of the most useful tools in studying manifolds with boundary is the "collaring theorem" which states that there is a neighborhood of the boundary which resembles a collar, that is, it is homeomorphic to the cartesian product of the boundary and an interval. In its topological form it is due to Brown [1962].

5.23   TOPOLOGICAL COLLARING THEOREM   If M is a topological manifold with boundary ∂M, then there exists a neighborhood W of ∂M in M such that W is homeomorphic to $\partial M \times [0, 1]$ in such a way that ∂M corresponds naturally with $\partial M \times 0$.

PROOF   See Appendix II.   □

If M is a manifold with boundary, define the "double" of M to be the manifold $\hat{M}$ formed by identifying two copies of M along ∂M.

Exercise 8   Show that $(M, \partial M)$ is orientable if and only if $\hat{M}$ is orientable.

Exercise 9   Show that if $(M, \partial M)$ is orientable, then ∂M is an orientable manifold without boundary. Is the converse true?

5.24   THEOREM   If $(M, \partial M)$ is a compact connected orientable n-manifold with orientation s, then there exists a unique fundamental class $z \in H_n(M, \partial M)$ such that for each $p \in M - \partial M$, $j_{p*}(z) = s(p)$. Furthermore, if $\Delta: H_n(M, \partial M) \to H_{n-1}(\partial M)$ is the connecting homomorphism, then $\Delta(z)$ is a fundamental class of ∂M, that is, it restricts to a fundamental class on each component of ∂M.

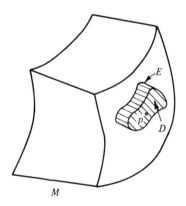

Figure 5.15

PROOF   Since $\hat{M}$ is orientable, there exists a fundamental class $\hat{z} \in H_n(\hat{M})$ such that $\hat{j}_{p*}(\hat{z}) = \hat{s}(p)$ for all $p \in \hat{M}$. Then we define z to be the image of $\hat{z}$ under the composition

$$H_n(\hat{M}) \to H_n(\hat{M}, \hat{M} - (M - \partial M)) \approx H_n(M, \partial M),$$

where the second homomorphism is the inverse of an excision isomorphism. This gives the existence of the desired fundamental class z.

Let D be a proper $(n - 1)$-ball in ∂M. If W is a collar for ∂M in M (Theorem 5.23), then under the homeomorphism $W \approx \partial M \times I$ there is an n-cell E corresponding to $D \times I$ (Figure 5.15). For any point p in the interior, D°, of the $(n - 1)$-cell D in ∂M we have the following diagram:

$$
\begin{array}{ccccc}
H_n(M, \partial M) & \xrightarrow{\ i_*\ } & H_n(M, M - E^\circ) & \xleftarrow[\approx]{\ k_*\ } & H_n(E, \partial E) \\
\downarrow{\scriptstyle \varDelta} & & \downarrow{\scriptstyle \varDelta} & & {\scriptstyle \approx}\downarrow{\scriptstyle \varDelta} \\
H_{n-1}(\partial M) & \xrightarrow{\ i_*'\ } & H_{n-1}(M - E^\circ) & \xleftarrow{\ k_*'\ } & H_{n-1}(\partial E) \\
\downarrow{\scriptstyle j_{p*}} & & \downarrow & & \downarrow{\scriptstyle \approx} \\
H_{n-1}(\partial M, \partial M - p) & \xrightarrow{\approx} & H_{n-1}(M - E^\circ, (M - E^\circ) - p) & \xleftarrow{\approx} & H_{n-1}(\partial E, \partial E - p)
\end{array}
$$

in which each rectangle commutes and the horizontal isomorphisms follow by excision.

If q is a point in E°, the factorization

$$
\begin{array}{ccc}
H_n(M, \partial M) & \xrightarrow{\ j_{q*}\ } & H_n(M, M - q) \\
{\scriptstyle i_*}\searrow & & \nearrow \\
& H_n(M, M - E^\circ) &
\end{array}
$$

and the fact that $j_{q*}(z)$ is the generator $s(q)$ imply that $i_*(z)$ is a generator for the infinite cyclic group $H_n(M, M - E^0)$. Thus, there exists a generator z' for $H_n(E, \partial E)$ such that $k_*(z') = i_*(z)$.

From the diagram $k_*'\varDelta(z') = i_*'\varDelta(z)$ and hence the images of $\varDelta(z)$ and $\varDelta(z')$ in the infinite cyclic group $H_{n-1}(M - E^\circ, (M - E^\circ) - p)$ must coincide. Since the image of $\varDelta(z')$ is a generator, so is the image of $\varDelta(z)$. Therefore, $j_{p*}(\varDelta(z))$ is a generator of $H_{n-1}(\partial M, \partial M - p)$ and $\varDelta(z)$ must be a fundamental class for ∂M.

To prove uniqueness, note first that if W is a collar for ∂M, then both M and $M - \partial M$ are homotopy equivalent to $M - W^\circ$. Thus

$$H_n(M) \approx H_n(M - \partial M) = 0$$

by Theorem 5.5. So suppose z and w are fundamental classes in $H_n(M, \partial M)$ corresponding to the orientation s. Since $\Delta(z)$ and $\Delta(w)$ are fundamental classes in $H_{n-1}(\partial M)$ corresponding to the orientation induced by s on ∂M, the restrictions of $\Delta(z)$ and $\Delta(w)$ to each component of ∂M must agree by Theorem 5.10. Thus, $z - w$ is in the kernel of Δ. By exactness and the fact that $H_n(M) = 0$ it follows that $z = w$. $\square$

5.25 POINCARÉ–LEFSCHETZ DUALITY THEOREM   Let $(M, \partial M)$ be a compact orientable n-manifold with fundamental class $z \in H_n(M, \partial M)$. Then the duality maps

$$D: \ H^k(M, \partial M) \to H_{n-k}(M) \quad \text{and} \quad D: \ H^k(M) \to H_{n-k}(M, \partial M)$$

given by taking the cap product with z are both isomorphisms.

PROOF   In $\hat{M}$ let M_1 and M_2 denote the two copies of M (Figure 5.16). There exists a two-sided collaring N of ∂M in $\hat{M}$. That is, N is homeomorphic to $\partial M \times I$, where $I = [-1, 1]$, with ∂M corresponding to $\partial M \times \{0\}$. Note that ∂N is homeomorphic to $\partial M \times \partial I$.

For $i = 1, 2$ consider the following diagram:

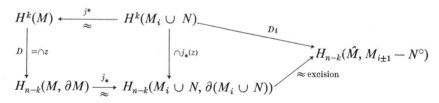

where D_i is defined to make the triangle commute. Since j is the inclusion map, the rectangle commutes by the naturality of the cap product;

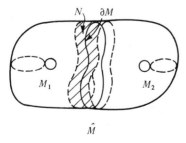

Figure 5.16

for any point $x \in M$. This implies the existence of a monomorphism

$$Z_p \oplus Z_p \oplus \text{Tor}(A, Z_p) \to Z_p,$$

which is impossible. Thus, $H_{n-1}(M)$ is free abelian. $\square$

It follows immediately from Lemma 5.26 that if M is a closed, connected, oriented n-manifold,

$$\text{Ext}(H_{n-1}(M), Z) = 0$$

so that

$$H^n(M) \approx \text{Hom}(H_n(M), Z),$$

which is infinite cyclic. If $z \in H_n(M)$ is the fundamental class corresponding to the given orientation, define $\alpha \in H^n(M)$ to be the "dual" of z in the sense that $\langle \alpha, z \rangle = 1$, so α is a generator for $H^n(M)$.

For any integer q define a pairing

$$H^q(M) \otimes H^{n-q}(M) \to Z$$

by sending $x \otimes y$ into the integer $\langle x \cup y, z \rangle$. Note that if $r \cdot x = 0$ in $H^q(M)$ for some integer $r \neq 0$, then $(r \cdot x) \cup y = r \cdot (x \cup y) = 0$; hence, $x \cup y = 0$ because $H^n(M)$ is infinite cyclic. Similarly $x \cup y = 0$ if y has finite order.

On the other hand, suppose that $x \in H^q(M)$ does not have finite order. From the universal coefficient theorem the homomorphism

$$H^q(M) \to \text{Hom}(H_q(M), Z),$$

sending x into the homomorphism $w \to \langle x, w \rangle$, must take x into a nontrivial homomorphism. Thus, there exists an element $w \in H_q(M)$ with $\langle x, w \rangle \neq 0$. Furthermore this is a split monomorphism, so that if x generates a direct summand of $H^q(M)$, then there exists an element $w \in H_q(M)$ with $\langle x, w \rangle = 1$.

Now by the Poincaré duality theorem there is an element $y \in H^{n-q}(M)$ with $y \cap z = w$. Then

$$\langle y \cup x, z \rangle = \langle x, y \cap z \rangle = \langle x, w \rangle \neq 0$$

and so $x \cup y \neq 0$. This completes the proof of the following:

5.27 PROPOSITION If M is a closed connected oriented n-manifold and $A_q \subseteq H^q(M)$ is the torsion subgroup, then there is a nonsingular dual pairing

$$H^q(M)/A_q \otimes H^{n-q}(M)/A_{n-q} \to Z. \quad \square$$

5.28 COROLLARY   If $a \in H^2(\mathbb{C}P(n))$ is a generator, then $a^k \in H^{2k}(\mathbb{C}P(n))$ is a generator for $1 \le k \le n$.

PROOF $\mathbb{C}P(n)$ is an orientable, compact, connected $2n$-manifold whose cohomology is given by

$$H^m(\mathbb{C}P(n)) \approx \begin{cases} Z & \text{for } m \text{ even,} \quad 0 \le m \le 2n \\ 0 & \text{otherwise.} \end{cases}$$

We prove the result by induction on n. It is obviously true for $n = 1$, so suppose it is true for $n - 1 \ge 1$, and consider the inclusion

$$i: \quad \mathbb{C}P(n-1) \subseteq \mathbb{C}P(n)$$

of the finite subcomplex which contains all cells of $\mathbb{C}P(n)$ except the one cell of dimension $2n$. From the exact sequence of the pair

$$\cdots \to H^{2k}(\mathbb{C}P(n), \mathbb{C}P(n-1)) \to H^{2k}(\mathbb{C}P(n)) \xrightarrow{i^*} H^{2k}(\mathbb{C}P(n-1))$$
$$\to H^{2k-1}(\mathbb{C}P(n), \mathbb{C}P(n-1)) \to \cdots$$

we see that i^* is an isomorphism for $2k < 2n$. Since $i^*(a)$ generates $H^2(\mathbb{C}P(n-1))$, the inductive hypothesis implies that $[i^*(a)]^k$ generates $H^{2k}(\mathbb{C}P(n-1))$ for all $k < n$. Now i^* is a ring homomorphism, so a^k must generate $H^{2k}(\mathbb{C}P(n))$ for $k < n$.

Finally, by Proposition 5.27 there is an element $b \in H^{2n-2}(\mathbb{C}P(n))$ with $a \cup b$ generating $H^{2n}(\mathbb{C}P(n))$. This b must generate $H^{2n-2}(\mathbb{C}P(n))$ so that $b = \pm a^{n-1}$. Therefore, $a \cup b = \pm a^n$ is a generator of $H^{2n}(\mathbb{C}P(n))$. This completes the proof. $\square$

Note that this completely describes the structure of the cohomology ring of $\mathbb{C}P(n)$.

5.29 COROLLARY $H^*(\mathbb{C}P(n))$ is a polynomial ring over the integers with one generator a in dimension two, subject to the relation $a^{n+1} = 0$. $\square$

Now let R be a field and M be a closed, connected, oriented manifold. As we observed previously

$$H^n(M; R) \approx \text{Hom}_R(H_n(M; R), R)$$

and

$$H^n(M; R) \approx R, \qquad H_n(M; R) \approx R.$$

Denote by $z_R \in H_n(M; R)$ a generator as an R-module. Then a slight variation of the Poincaré duality theorem states that the homomorphism

$$H^q(M; R) \to H_{n-q}(M; R),$$

given by sending a into $a \cap z_R$, is an isomorphism. The technique of Proposition 5.27 may now be used to prove the following.

5.30 PROPOSITION The pairing $H^q(M; R) \otimes H^{n-q}(M; R) \to R$ given by sending $x \otimes y$ into $\langle x \cup y, z_R \rangle \in R$ is a nonsingular dual pairing. $\square$

5.31 COROLLARY If M is a closed, connected n-manifold, then

$$H^q(M; Z_2) \otimes H^{n-q}(M; Z_2) \to Z_2$$

is a nonsingular dual pairing. $\square$

5.32 COROLLARY If $a \in H^1(\mathbb{R}P(n); Z_2)$ is a generator, then a^k generates $H^k(\mathbb{R}P(n); Z_2)$ for $1 \le k \le n$. Thus, $H^*(\mathbb{R}P(n); Z_2)$ is a polynomial ring over Z_2 with one generator a in dimension one, subject to the relation $a^{n+1} = 0$. $\square$

Similar arguments may be used to compute the cohomology ring of quaternionic projective space $H\mathbb{P}(n)$.

5.33 COROLLARY $H^*(H\mathbb{P}(n))$ is a polynomial ring over the integers with one generator a in dimension four subject to the relation $a^{n+1} = 0$. $\square$

With these results we may now establish the existence of certain maps having odd Hopf invariant (see Chapter 4).

5.34 COROLLARY There exist maps $S^3 \to S^2$, $S^7 \to S^4$, and $S^{15} \to S^8$ having Hopf invariant 1.

PROOF Let $f: S^3 \to S^2$ be the Hopf map of Chapter 2. Recall that S_f^2 is homeomorphic to $\mathbb{C}P(2)$. So if b is a generator of $H^2(S_f^2)$ and a is a generator of $H^4(S_f^2)$, then $b^2 = \pm a$ by Corollary 5.29. Thus, $H(f) = \pm 1$. Now the results of Exercise 8 (i) of Chapter 4 indicate how to alter f, if necessary, to give a map with Hopf invariant 1.

The cases $S^7 \to S^4$ and $S^{15} \to S^8$ follow by applying the same approach to the quaternions and the Cayley numbers, respectively. ☐

In order to develop some further applications we must introduce some basic facts about bilinear forms. Let V be a real vector space of finite dimension. A bilinear form

$$\Phi: \quad V \times V \to \mathbb{R}$$

is *nonsingular* if $\Phi(x, y) = 0$ for all y in V implies $x = 0$.

Exercise 11 Show that this is equivalent to requiring that $\Phi(x, y) = 0$ for all x in V imply $y = 0$.

The form Φ is *symmetric* if $\Phi(x, y) = \Phi(y, x)$; it is *antisymmetric* if $\Phi(x, y) = -\Phi(y, x)$, for all x and y in V.

Example Let $V = \mathbb{R}^2$ and denote its points by (x, y). Define

$$\Phi((x, y), (x', y')) = \det\begin{pmatrix} x & y \\ x' & y' \end{pmatrix}.$$

This is a nonsingular, antisymmetric bilinear form.

Given any bilinear form Φ on $V \times V$ we may write Φ uniquely as $\Phi = \Phi' + \Phi''$, where Φ' is symmetric and Φ'' is antisymmetric. To see this we set

$$\Phi'(x, y) = \tfrac{1}{2}[\Phi(x, y) + \Phi(y, x)]$$

and

$$\Phi''(x, y) = \tfrac{1}{2}[\Phi(x, y) - \Phi(y, x)].$$

Let Φ be antisymmetric on $V \times V$. Then $\Phi(x, x) = 0$ for all $x \in V$. If $x_1 \in V$ has $\Phi(x_1, y) \neq 0$ for some y, then obviously there exists an element y_1 in V with

$$\Phi(x_1, y_1) = 1.$$

Define V_1 to be the subspace of V given by

$$V_1 = \{x \in V \mid \Phi(x, x_1) = 0 \text{ and } \Phi(x, y_1) = 0\}.$$

This linear subspace may be identified with the kernel of the linear transformation

$$\theta: \quad V \to \mathbb{R}^2$$

given by $\theta(x) = (\Phi(x, x_1), \Phi(x, y_1))$. Note that since $\theta(x_1) = (0, 1)$ and $\theta(y_1) = (-1, 0)$, the transformation is an epimorphism. Therefore, the dimension of V_1 is equal to dimension $V - 2$. By repeating this process using the subspace V_1 to produce a subspace V_2, and so forth, we will eventually either exhaust V or reach a subspace with the property that the product of any pair of its vectors is zero.

Thus, there is a basis for V of the form

$$x_1, \quad y_1, \quad x_2, \quad y_2, \ldots, x_k, \quad y_k, \quad z_1, \ldots, z_s$$

for which $\Phi(x_i, y_i) = 1 = -\Phi(y_i, x_i)$, and any other pair of basis vectors have product zero.

5.35   LEMMA   If $\Phi: V \times V \to R$ is nonsingular and antisymmetric, then the dimension of V is even.

PROOF   This follows from the above, since $s = 0$. $\square$

5.36   COROLLARY   If M is a closed, oriented manifold of dimension $4k + 2$, then $\chi(M)$ is even.

PROOF   Recall that the Euler characteristic is given by

$$\chi(M) = \sum_{i=0}^{4k+2} (-1)^i \dim H_i(M; R).$$

By the universal coefficient theorem this may also be expressed as the sum

$$\chi(M) = \sum_{i=0}^{4k+2} (-1)^i \dim H^i(M; R).$$

Since M is closed and oriented, the Poincaré duality theorem implies

$$H^i(M; R) \approx H_{4k+2-i}(M; R) \approx \mathrm{Hom}(H^{4k+2-i}(M; R), R).$$

Therefore, $\dim H^i(M; R) = \dim H^{4k+2-i}(M; R)$. As a result the entries in the second sum are paired up, except in the middle dimension, so that

$$\sum_{i=2k+1} (-1)^i \dim H^i(M; R)$$

is even.

Finally note that there is a bilinear form

$$\Phi: \quad H^{2k+1}(M; \mathbb{R}) \times H^{2k+1}(M; \mathbb{R}) \to \mathbb{R}$$

defined by $\Phi(x, y) = \langle x \cup y, z_R \rangle \in \mathbb{R}$. By Proposition 5.30 this is non-singular, and since x and y are both odd dimensional, it is antisymmetric. Thus, by Lemma 5.35 the dimension of $H^{2k+1}(M; \mathbb{R})$ is even and $\chi(M)$ is even. $\square$

5.37 COROLLARY If M is a closed manifold of dimension $2k + 1$, then $\chi(M) = 0$.

PROOF Since $H_i(M)$ is a finitely generated abelian group for each i, we may write

$$H_i(M) \approx A_i \oplus B_i \oplus C_i,$$

where A_i is free abelian of rank r_i, B_i is a direct sum of s_i cyclic groups of order a power of two, and C_i is a direct sum of cyclic groups of odd order. Note that

$$\chi(M) = \sum_{i=0}^{2k+1} (-1)^i r_i.$$

By the universal coefficient theorem

$$\dim H_i(M; Z_2) = \dim(H_i(M) \otimes Z_2) + \dim(\text{Tor}(H_{i-1}(M), Z_2))$$
$$= (r + s_i) + (s_{i-1}).$$

Thus

$$\sum_{i=0}^{2k+1} (-1)^i \dim H_i(M; Z_2) = \sum_{i=0}^{2k+1} (-1)^i [r_i + s_i + s_{i-1}]$$
$$= \sum_{i=0}^{2k+1} (-1)^i r_i = \chi(M).$$

On the other hand, by the Poincaré duality theorem

$$H_i(M; Z_2) \approx H^{2k+1-i}(M; Z_2) \approx \text{Hom}(H_{2k+1-i}(M; Z_2), Z_2)$$

so that

$$\dim H_i(M; Z_2) = \dim H_{2k+1-i}(M; Z_2).$$

Since i and $2k + 1 - i$ have different parity, these appear in the sum with opposite signs. Therefore

$$\chi(M) = \sum_{i=0}^{2k+1} (-1)^i \dim H_i(M; Z_2) = 0. \qquad \square$$

NOTE This result is obviously false for even-dimensional manifolds since $\chi(S^{2n}) = 2$, $\chi(\mathbb{R}P(2n)) = 1$, and $\chi(\mathbb{C}P(n)) = n + 1$. The vanishing of the Euler characteristic is a useful fact in differential geometry, as is seen in the following basic theorem: a closed differentiable manifold M admits a nonzero vector field if and only if $\chi(M) = 0$. Thus, Corollary 5.37 implies that any odd-dimensional, closed, differentiable manifold admits a nonzero vector field. We shall return to this subject in Chapter 6.

Now suppose that $\Phi: V \times V \to \mathbb{R}$ is symmetric. Then since

$$\Phi(x + y, x + y) = \Phi(x, x) + 2\Phi(x, y) + \Phi(y, y)$$

or

$$\Phi(x, y) = \tfrac{1}{2}[\Phi(x + y, x + y) - \Phi(x, x) - \Phi(y, y)],$$

it follows that if Φ is nontrivial, there exists an $x_1 \in V$ with $\Phi(x_1, x_1) \neq 0$. We may as well assume $\Phi(x_1, x_1) = \pm 1$. Consider the homomorphism

$$\alpha: \quad V \to \mathbb{R}$$

given by $\alpha(x) = \Phi(x, x_1)$. This is an epimorphism since $\alpha(x_1) = \pm 1$, so if V_1 is the kernel of α, the dimension of V_1 is one less than the dimension of V. By applying the same analysis to V_1 to give an element x_2, and continuing the process, we produce a basis for V which may be renumbered so as to have the form $x_1, \ldots, x_r, x_{r+1}, \ldots, x_{r+s}, x_{r+s+1}, \ldots, x_{r+s+t}$, where

$$\Phi(x_i, x_i) = \begin{cases} 1 & \text{if } 1 \leq i \leq r \\ -1 & \text{if } r < i \leq r + s \\ 0 & \text{if } r + s < i \leq r + s + t, \end{cases}$$

and any other pair has product zero.

Exercise 12 Show that the numbers r and s are invariants of the symmetric form Φ; that is, that they are independent of the various choices made.

The *signature* of a symmetric form Φ is the integer $r - s$. If Φ is an arbitrary bilinear form, then we write $\Phi = \Phi' + \Phi''$ with Φ' symmetric and

Φ'' antisymmetric. We then define the signature of Φ to be the signature of Φ'.

Let M be a closed, oriented n-manifold. Define the *index* of M, denoted $\tau(M)$, as follows:

(i) $\tau(M) = 0$ if $n \neq 4k$ for some integer k;

(ii) if $n = 4k$, let $\tau(M)$ be the signature of the nonsingular symmetric bilinear form

$$\Phi: \quad H^{2k}(M; \mathbb{R}) \times H^{2k}(M; \mathbb{R}) \to \mathbb{R}.$$

Exercise 13 Let M be a closed, connected oriented $4k$-manifold. Define a bilinear form

$$\Psi: \quad H^*(M; \mathbb{R}) \times H^*(M; \mathbb{R}) \to \mathbb{R}$$

by

$$\Psi(x, y) = \langle (x \cup y)_{4k}, z_R \rangle \in \mathbb{R},$$

where $(x \cup y)_{4k}$ is the $4k$-dimensional component of $x \cup y$. Then show that the signature of Ψ is $\tau(M)$.

Exercise 14 Let M_1 and M_2 be disjoint, closed, connected oriented manifolds.

(a) Show that the manifold $M_1 \times M_2$ may be oriented in such a way that

$$\tau(M_1 \times M_2) = \tau(M_1) \cdot \tau(M_2).$$

(b) If M_1 and M_2 have the same dimension, show that

$$\tau(M_1 \cup M_2) = \tau(M_1) + \tau(M_2).$$

Note that a change in the orientation of a manifold merely changes the sign of its index. It is evident that the index of $\mathbb{C}P(2k)$ is ± 1, depending on the choice of orientation. Thus, it follows from the above exercise that there exist $4k$-dimensional manifolds of arbitrary index for all values of k.

The final question we would like to consider is the following: given a closed topological manifold M, when does there exist a compact manifold W with $M = \partial W$? Of course we must require that W be compact since M is always the boundary of $M \times [0, 1)$. Our first result gives a necessary condition for M to be such a boundary.

5.38   THEOREM   If W is a compact topological manifold with $\partial W = M$, then $\chi(M)$ is even.

PROOF   If the dimension of M is odd, then $\chi(M) = 0$ by Corollary 5.37. Thus we assume that the dimension of M is even so that the dimension of W is odd. Now consider the manifold $W \times I$ (see Exercise 15), where $I = [0, 1]$. We have

$$\partial(W \times I) = M \times I \cup W \times \partial I = M \times I \cup W \times \{0\} \cup W \times \{1\}.$$

Define $U = \partial(W \times I) - W \times \{1\}$ and $V = \partial(W \times I) - W \times \{0\}$. Note that U and V are open subsets of $\partial(W \times I)$ and W, U, and V all have the same homotopy type, whereas $U \cap V$ has the homotopy type of M.

The Mayer–Vietoris sequence for U and V becomes

$$H_{i+1}(\partial(W \times I)) \xrightarrow{h_{i+1}} H_i(M) \xrightarrow{f_i} H_i(W) \oplus H_i(W) \xrightarrow{g_i} H_i(\partial(W \times I)),$$

where each group is finitely generated and zero in dimensions greater than the dimension of W.

From the exactness we see that

$$\mathrm{rank}(H_i(M)) = \mathrm{rank}(\mathrm{image}\ h_{i+1}) + \mathrm{rank}(\mathrm{image}\ f_i),$$
$$\mathrm{rank}(H_i(W) \oplus H_i(W)) = \mathrm{rank}(\mathrm{image}\ f_i) + \mathrm{rank}(\mathrm{image}\ g_i),$$
$$\mathrm{rank}(H_i(\partial(W \times I))) = \mathrm{rank}(\mathrm{image}\ g_i) + \mathrm{rank}(\mathrm{image}\ h_i).$$

By multiple cancellations it follows that

$$\sum (-1)^i \, \mathrm{rank}(H_i(M)) - \sum (-1)^i \, \mathrm{rank}(H_i(W) \oplus H_i(W))$$
$$+ \sum (-1)^i \, \mathrm{rank}(H_i(\partial(W \times I))) = 0.$$

Since $\partial(W \times I)$ is an odd-dimensional, closed manifold, we have $\chi(\partial(W \times I)) = 0$ by Corollary 5.37. Therefore

$$\chi(M) = 2 \cdot \sum (-1)^i \, \mathrm{rank}\ H_i(W) = 2 \cdot \chi(W). \qquad \square$$

Exercise 15   Suppose M_1 and M_2 are topological manifolds. Then show that $M_1 \times M_2$ is a topological manifold with

$$\partial(M_1 \times M_2) = (\partial M_1) \times M_2 \cup M_1 \times (\partial M_2).$$

Note that as an immediate consequence of Theorem 5.38 we have many manifolds which are not boundaries of compact manifolds, for example, $\mathbb{R}P(2k)$ and $\mathbb{C}P(2k)$.

A necessary condition for a closed manifold M to bound a compact oriented manifold is that the index of M be zero. In order to prove this we will need the following:

5.39   LEMMA   Suppose Φ is a symmetric, nonsingular bilinear form on a vector space V of dimension $2n$, and $\{x_1, \ldots, x_n\}$ is a linearly independent set in V such that $\Phi(\sum a_i x_i, \sum b_j x_j) = 0$ for any coefficients $a_1, \ldots, a_n$, $b_1, \ldots, b_n$. Then the signature of Φ is zero.

PROOF   In the decomposition described previously, it is evident that t must be zero since Φ is nonsingular. We must show that $r = s = n$. We will prove inductively that $r \geq n$; a similar argument establishes $s \geq n$, from which the conclusion follows.

For $n = 1$, there exists an element y_1 in V with $\Phi(x_1, y_1) \neq 0$. Then

$$\Phi(y_1 + ax_1, y_1 + ax_1) = \Phi(y_1, y_1) + 2a\Phi(x_1, y_1)$$

so by choosing

$$a = \frac{1 - \Phi(y_1, y_1)}{2\Phi(x_1, y_1)}$$

we have $\Phi(y_1 + ax_1, y_1 + ax_1) = 1$ and $r \geq 1 = n$.

Now suppose the assertion is true for vector spaces of dimension $2(n - 1)$. Define a homomorphism

$$\Theta: \quad V \to \mathbb{R}^n$$

by $\Theta(z) = (\Phi(x_1, z), \ldots, \Phi(x_n, z))$. If Θ is not an epimorphism, then the dimension of the kernel of Θ is $\geq n + 1$. On the other hand, we may extend the linearly independent set to a basis $\{x_1, \ldots, x_n, \omega_1, \ldots, \omega_n\}$ for V and define

$$\Theta': \quad V \to \mathbb{R}^n$$

by $\Theta'(z) = (\Phi(\omega_1, z), \ldots, \Phi(\omega_n, z))$. The dimension of the kernel of Θ' is $\geq n$; hence

$$\ker \Theta \cap \ker \Theta' \neq 0.$$

But this cannot happen since Φ is nonsingular, so Θ must be an epimorphism.

Let $y_1 \in \Theta^{-1}(1, 0, \ldots, 0)$. As before, there exists a real number a with $\Phi(y_1 + ax_1, y_1 + ax_1) = 1$. Now define

$$\Psi: \quad V \to \mathbb{R}^2$$

by $\Psi(z) = (\Phi(x_1, z), \Phi(y_1, z))$ and note that Ψ is an epimorphism. If V' is the kernel of Ψ, then the restriction of Φ to V' is a nonsingular form and $\{x_2, \ldots, x_n\}$ is a linearly independent set in V' satisfying the hypothesis. Thus, by the inductive hypothesis there exist vectors $q_2, \ldots, q_n$ in V' with $\Phi(q_i, q_j) = \delta_{ij}$. The collection $y_1 + ax_1, q_2, \ldots, q_n$ then shows that $r \geq n$. $\square$

5.40 THEOREM If M is a compact oriented $(4n + 1)$-manifold with boundary, then the index of ∂M is zero.

PROOF Denote by Φ the symmetric nonsingular bilinear form on $H^{2n}(\partial M; \mathbb{R})$. We will show that the signature of Φ is zero by proving that the image of

$$j^*: \quad H^{2n}(M; \mathbb{R}) \to H^{2n}(\partial M; \mathbb{R})$$

is a subspace of half the dimension of $H^{2n}(\partial M; \mathbb{R})$ on which Φ is identically zero, where $j: \partial M \to M$ is the inclusion.

Let $z_{\mathbb{R}} \in H_{4n+1}(M, \partial M; \mathbb{R})$ be a fundamental class and take $\Delta z_{\mathbb{R}} \in H_{4n}(\partial M; \mathbb{R})$ to be the fundamental class given by the image of $z_{\mathbb{R}}$ under the connecting homomorphism. If $j^*(\alpha)$ and $j^*(\beta)$ are two elements of $H^{2n}(\partial M; \mathbb{R})$ in the image of j^*, then

$$\Phi(j^*(\alpha), j^*(\beta)) = \langle j^*(\alpha) \cup j^*(\beta), \Delta z_{\mathbb{R}} \rangle$$
$$= \langle j^*(\alpha \cup \beta), \Delta z_{\mathbb{R}} \rangle$$
$$= \langle \alpha \cup \beta, j_* \Delta z_{\mathbb{R}} \rangle$$
$$= 0.$$

So Φ is identically zero on the image of j^*.

Consider the commutative diagram

$$
\begin{array}{ccc}
H^{2n}(M; \mathbb{R}) & \xrightarrow{j^*} & H^{2n}(\partial M; \mathbb{R}) \\
\approx \downarrow D' & & \approx \downarrow D \\
H_{2n+1}(M, \partial M; \mathbb{R}) & \xrightarrow{\Delta} & H_{2n}(\partial M; \mathbb{R})
\end{array}
$$

as in the proof of Theorem 5.25, where D and D' are Poincaré duality isomorphisms. Then since $D(j^*(\alpha)) = \Delta(D'(\alpha))$, it follows that the image of j^* is isomorphic to the image of Δ. Thus, the dimension of the image of j^* is the same as the dimension of the kernel of j_*.

On the other hand, since R is a field, the universal coefficient theorem gives a commutative diagram

$$
\begin{array}{ccc}
H^{2n}(M;R) & \xrightarrow{\approx} & \mathrm{Hom}(H_{2n}(M;R), R) \\
\Big\downarrow{\scriptstyle j^*} & & \Big\downarrow{\scriptstyle (j_*)^{\#}} \\
H^{2n}(\partial M;R) & \xrightarrow{\approx} & \mathrm{Hom}(H_{2n}(\partial M;R), R)
\end{array}
$$

in which the horizontal maps are isomorphisms. Then it is easily checked that the dimension of the image of j^* is equal to the dimension of the image of j_*.

Putting these together we have

$$
\begin{aligned}
2 \cdot \dim \mathrm{im}\, j^* &= \dim \ker j_* + \dim \mathrm{im}\, j_* \\
&= \dim H_{2n}(\partial M;R) \\
&= \dim H^{2n}(\partial M;R).
\end{aligned}
$$

Thus, the image of j^* is a subspace of $H^{2n}(\partial M;R)$ of half the dimension. It follows from Lemma 5.39 that the index of ∂M is zero.   □

Note that Theorems 5.38 and 5.40 give certain necessary conditions for closed manifolds to be boundaries of compact manifolds of one dimension higher. These conditions are more closely related than may be readily apparent.

5.41 PROPOSITION   If M^n is a closed oriented manifold, then

$$
\tau(M) \equiv \chi(M) \bmod 2.
$$

PROOF   This is clear if the dimension of M is odd since $\chi(M) = 0 = \tau(M)$. If $\dim M \equiv 2 \bmod 4$, then by Corollary 5.36 $\chi(M)$ is even, hence congruent to $\tau(M) \bmod 2$. If the dimension of M is $4k$, then $\chi(M) \equiv \dim H^{2k}(M;R)$ mod 2. On the other hand, $\tau(M) = r - s$, where $r + s = \dim H^{2k}(M;R)$. Thus

$$
\chi(M) - \tau(M) \equiv 2s \equiv 0 \bmod 2.   \qquad □
$$

From considering such examples as S^{2n} or $CP(n)$ it is apparent that the congruence in Proposition 5.41 cannot be replaced by equality.

"Index" invariants of this type for manifolds are very important in algebraic and differential topology. Of particular interest is their connection with analysis, which arises from the analytical interpretation of cohomology groups via the theory of Hodge and de Rham. Much significant progress has been made in recent years in relating geometrical invariants to such analytical invariants as the indices of differential operators.

The results of Theorems 5.38 and 5.40 introduce us to another area of considerable current interest. Let M^n be a closed, oriented manifold. M^n is said to *bound* if there exists a compact, oriented manifold W^{n+1} and an orientation-preserving homeomorphism of M^n onto ∂W^{n+1}. Note that it is essential to require W^{n+1} to be compact, as M^n will always bound $M^n \times [0, 1)$ if $[0, 1)$ is properly oriented.

Two closed, oriented n-manifolds M_1^n and M_2^n are *oriented cobordant*, $M_1^n \sim M_2^n$, if the manifold given by the disjoint union $M_1^n \cup -M_2^n$ bounds a compact $(n + 1)$-manifold W^{n+1}, where $-M_2^n$ is the manifold given by reversing the orientation on M_2^n. The manifold W^{n+1} is called a *cobordism* between M_1^n and M_2^n.

This defines an equivalence relation on the class of closed oriented n-manifolds. To see this, note that $M^n \sim M^n$ because $M^n \cup -M^n$ is homeomorphic to the boundary of $M^n \times [0, 1]$ by an orientation-preserving homeomorphism. To establish transitivity we glue two cobordisms together (Figure 5.17). That is, if $\partial W^{n+1} = M_1^n \cup -M_2^n$ and $\partial V^{n+1} = M_2^n \cup -M_3^n$ then by identifying W^{n+1} and V^{n+1} along the common copy of M_2^n we get a compact oriented manifold with boundary oriented homeomorphic to $M_1^n \cup -M_3^n$.

Let $[M^n]$ be the equivalence class represented by M^n. Denote the set of equivalence classes by $\mathfrak{N}_n^{\text{STOP}}$. We define an additive operation in $\mathfrak{N}_n^{\text{STOP}}$ by setting $[M_1^n] + [M_2^n] = [M_1^n \cup M_2^n]$, the equivalence class of the disjoint union. This gives $\mathfrak{N}_n^{\text{STOP}}$ the structure of an abelian group in which $-[M^n] = [-M^n]$ and the additive identity is the equivalence class of those

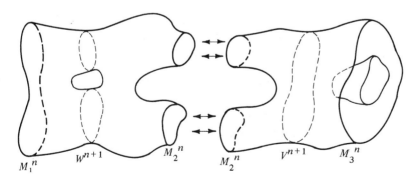

Figure 5.17

manifolds which bound. The graded group

$$\mathfrak{N}_*^{\text{STOP}} = \sum_{n=0}^{\infty} \mathfrak{N}_n^{\text{STOP}}$$

may be given the structure of a commutative graded ring by defining $[M_1{}^n] \cdot [M_2{}^m] = [M_1{}^n \times M_2{}^m]$, the unit being the class of a positively oriented point in $\mathfrak{N}_0^{\text{STOP}}$. $\mathfrak{N}_*^{\text{STOP}}$ is the *oriented topological cobordism ring.*

If in the previous discussion we omit all references to orientation, there result the *unoriented* topological cobordism groups $\mathfrak{N}_n^{\text{TOP}}$. Denoting the unoriented equivalence class of the closed manifold M^n by $[M^n]_2$, it is apparent that $2 \cdot [M^n]_2 = 0$ since $M^n \cup M^n$ bounds $M^n \times [0, 1]$. Thus the *unoriented topological cobordism ring* $\mathfrak{N}_*^{\text{TOP}}$ becomes a Z_2-algebra.

Define mappings $\Psi \colon \mathfrak{N}_*^{\text{STOP}} \to Z$ and $\Psi_2 \colon \mathfrak{N}_*^{\text{TOP}} \to Z_2$ by $\Psi([M^n]) = \tau(M^n)$, the index of M^n, and $\Psi_2([M^n]_2) = \chi(M^n)$ reduced mod 2. By Theorems 5.38 and 5.40 these are well-defined functions of the respective cobordism classes. Furthermore, since both invariants are additive over disjoint unions and multiplicative over cartesian products, Ψ and Ψ_2 are ring homomorphisms.

A closed, oriented 0-manifold is a finite collection of points, each given a positive or negative orientation. It bounds if and only if the "algebraic" sum of the points is zero. Similarly an unoriented 0-manifold bounds if and only if it consists of an even number of points. Thus, $\Psi \colon \mathfrak{N}_0^{\text{STOP}} \to Z$ and $\Psi_2 \colon \mathfrak{N}_0^{\text{TOP}} \to Z_2$ are both isomorphisms.

Note that since any closed 1-manifold is homeomorphic to a finite disjoint union of circles, it follows immediately that $\mathfrak{N}_1^{\text{TOP}} = 0 = \mathfrak{N}_1^{\text{STOP}}$.

Exercise 16 From the classification of closed 2-manifolds, compute $\mathfrak{N}_2^{\text{TOP}}$ and $\mathfrak{N}_2^{\text{STOP}}$.

It is evident that since each $RP(2n)$ has Euler characteristic equal to one, $\Psi_2 \colon \mathfrak{N}_{2n}^{\text{TOP}} \to Z_2$ is an epimorphism for each n. Similarly each $CP(2n)$ has index ± 1 so that $\Psi \colon \mathfrak{N}_{4n}^{\text{STOP}} \to Z$ is an epimorphism for each n.

The structure of these rings has remained a mystery for some time. Recently the ring $\mathfrak{N}_*^{\text{TOP}}$ has been determined in all dimensions $\neq 4$ by Brumfiel *et al.* [1971], using results of Kirby and Siebenmann [1969]. As an excellent further reference in this area we recommend Stong [1968].

chapter 6

FIXED-POINT THEORY

We are interested in studying the behavior of continuous functions on manifolds with particular interest in detecting the presence or absence of fixed points or coincidences. This is a classical problem, so it may prove enlightening to take a brief look at some of its early development.

During the 1880s, Poincaré studied vector distributions on surfaces. For an isolated singularity of such a distribution he assigned an *index* which was an integer (positive, negative, or zero). A vector distribution may be interpreted as a map of the surface to itself by translating a point via the vector based at that point. Here the fixed points of the map are the singularities of the distribution. Thus, summing the indices of the isolated singularities was the first step toward "algebraically" counting the fixed points of a map. Poincaré proved that if the surface is orientable of genus p and the distribution has only isolated singularities, then the sum of the indices is $2 - 2p$.

At the beginning of the twentieth century, Brouwer defined the *degree* of a mapping between n-manifolds. This allowed him to prove his fixed point theorem for mappings of D^n as well as to extend from 2 to n dimensions the definition of the index given by Poincaré. One of his important results was: if f and g are homotopic mappings of an n-manifold to itself and both f and g have only a finite number of fixed points, then the sum of the indices of the fixed points is the same for both functions. Since every mapping can be deformed into one with only a finite number of fixed points, this produces a homotopy invariant for the "algebraic" number of fixed points.

In 1923 Lefschetz published the first version of his fixed-point formula. Let M be a closed manifold and $f: M \to M$ be a map. Then for each k there is the induced homomorphism on homology with rational coefficients

$$f_k: \quad H_k(M; \mathcal{Q}) \to H_k(M; \mathcal{Q}).$$

For each k we may choose a basis for the finite-dimensional, rational vector space $H_k(M; \mathcal{Q})$ and write f_k as a matrix with respect to this basis. Denote by $\mathrm{tr}(f_k)$ the trace of this matrix. If we define the *Lefschetz number* of f by

$$L(f) = \sum_{k=0}^{\infty} (-1)^k \, \mathrm{tr}(f_k),$$

then $L(f)$ is independent of the choices involved and hence is a well-defined, rational-valued function of f. It is evident that $L(f)$ depends only on the homotopy class of f.

To see how this is connected with the earlier work of Brouwer, consider the case of a closed orientable manifold. Lefschetz proved the following: for each $\varepsilon > 0$ there is an ε-approximation g to f (here we are assuming a metric on the manifold) such that (i) g has only a finite number of fixed points, and (ii) for each fixed point x of g, g takes some neighborhood of x homeomorphically onto some other neighborhood of x. If $x_1, \ldots, x_m$ are the fixed points of g, denote by $a_1, \ldots, a_m$ the local degrees of g at these points in the sense of Brouwer. Then Lefschetz showed that

$$L(g) = \sum_{i=1}^{m} a_i.$$

Now for ε small, f and g are homotopic; hence, $f_* = g_*$ and we have

$$L(f) = \sum_{i=1}^{m} a_i.$$

This implies that $L(f)$ is always an integer and leads to the celebrated *Lefschetz fixed-point theorem*: if $L(f) \neq 0$, then f has a fixed point.

The idea of the proof is as follows: in the product space $M \times M$ consider the diagonal $\Delta(M)$ (Figure 6.1). Denote by $G(g)$ the graph of the function g. The points of $\Delta(M) \cap G(g)$ correspond to the fixed points of g. The previously stated process of approximating f by g corresponds to making slight changes in $G(f)$ in order to put it into "general position" with respect to $\Delta(M)$. Here we see that the a_i have the proper interpretation, determined by that particular intersection of the graph with the diagonal. Considering $\Delta(M)$ and $G(g)$ as n-dimensional chains in $M \times M$, Lefschetz computed their intersection number and showed it to be the trace formula.

As a special case we may take f to be the identity map. Then $L(f) = \chi(M)$, the Euler characteristic of M. If M is a connected, differentiable manifold which admits a nonzero vector field, we may interpret this as before as a map homotopic to the identity but having no fixed points. Thus, $L(f) = 0$ and this implies $\chi(M) = 0$. The classical theorem of Hopf is the converse of this, that is, that if $\chi(M) = 0$ then M admits a nonzero vector field.

Generalizing the previous, if f and $g\colon M_1 \to M_2$ are maps between closed oriented n-manifolds, a *coincidence* of f and g is a point $x \in M_1$ such that $f(x) = g(x)$. Geometrically, if $G(f)$ and $G(g)$ are the graphs of the respective functions in $M_1 \times M_2$, their points of intersection correspond to the coincidences.

From the diagram

$$H_q(M_1; \mathcal{Q}) \xrightarrow{\ f_*\ } H_q(M_2; \mathcal{Q})$$
$$\approx \Big\uparrow \mu \qquad\qquad \approx \Big\uparrow \nu$$
$$H^{n-q}(M_1; \mathcal{Q}) \xleftarrow{\ g^*\ } H^{n-q}(M_2; \mathcal{Q})$$

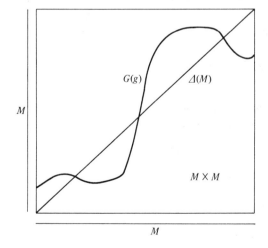

$G(g)$ $\Delta(M)$

$M \times M$

M

M

Figure 6.1

where the vertical homomorphisms are Poincaré duality isomorphisms, we define

$$\Theta_q: \quad H_q(M_1; \mathcal{Q}) \to H_q(M_1; \mathcal{Q})$$

to be $\Theta_q = \mu g^* \nu^{-1} f_*$. Then the *coincidence number* of f and g is given by

$$L(f, g) = \sum_{q=0}^{n} (-1)^q \operatorname{tr}(\Theta_q).$$

As before, $L(f, g)$ is the intersection number of $G(f)$ and $G(g)$, so if $L(f, g) \neq 0$, then f and g have a coincidence. Note that if $M_1 = M_2$ and g is the identity, then $L(f, g) = L(f)$.

In this chapter we will prove these major results in the framework of the previous chapters. We will do this by first defining the coincidence index and the fixed point index and establishing their basic properties. By introducing certain characteristic cohomology classes we establish the link between these indices and the corresponding coincidence numbers and Lefschetz numbers. In the process we encounter the Euler class and show that when evaluated on the fundamental class it yields the Euler characteristic. The principal tools used are the Poincaré duality theorem and the Thom isomorphism theorem. We close with some applications and observations.

It should be pointed out the spaces we consider, closed, oriented manifolds, could be made much more general. Similar techniques may be applied in the nonorientable case by using twisted coefficients. Many of the theorems are valid for such spaces as euclidean neighborhood retracts [Dold, 1965]. We impose these restrictions both for the purpose of continuity with the previous material and so that the reader may easily grasp the fundamental ideas involved. Many of the techniques of this chapter have been evolved from the excellent papers of Dold [1965] and Samelson [1965].

Let M_1 and M_2 be closed, connected, oriented n-manifolds with fundamental classes $z_i \in H_n(M_i)$, and corresponding Thom classes

$$U_i \in H^n(M_i \times M_i, \, M_i \times M_i - \Delta(M_i)), \qquad i = 1, 2.$$

Suppose W is an open set in M_1 and

$$f, g: \quad W \to M_2$$

are maps for which the coincidence set $C = \{x \in W \mid f(x) = g(x)\}$ is a compact subset of W.

By the normality of M_1 there exists an open set V in M_1 with $C \subseteq V \subseteq \bar{V} \subseteq W$. Define the *coincidence index* of the pair (f, g) on W to be the integer $I_{f,g}^W$ given by the image of the fundamental class z_1 under the composition

$$H_n(M_1) \to H_n(M_1, M_1 - V) \xrightarrow[\approx]{\text{excision}} H_n(W, W - V)$$

$$\xrightarrow{(f,g)_*} H_n(M_2 \times M_2, M_2 \times M_2 - \Delta(M_2)) \approx Z.$$

Here the map $(f, g): W \to M_2 \times M_2$ is given by $(f, g)(x) = (f(x), g(x))$, and the identification

$$H_n(M_2 \times M_2, M_2 \times M_2 - \Delta(M_2)) \approx Z$$

is given by sending a class α into the integer $\langle U_2, \alpha \rangle$. That this is an isomorphism follows from the fact in Equation (5.19) that for $p \in M_2$, $\langle U_2, l_{p*}(s(p)) \rangle = 1$.

It must first be shown that this definition is independent of the choice of the open set V. Suppose V' is another open set with $C \subseteq V' \subseteq \bar{V}' \subseteq W$. Then consider the following diagram:

Here, as in Chapter 5, $M_2^\times$ denotes the pair $(M_2 \times M_2, M_2 \times M_2 - \Delta(M_2))$. Since each triangle and rectangle commutes, the images of z_1 across the top and across the bottom must be the same. This shows that $I_{f,g}^W$ is independent of the choice of V.

Exercise 1 Let W' be another open set in M_1 and f' and $g': W' \to M_2$ be maps such that $f = f'$ and $g = g'$ on $W \cap W'$ and

$$C' = \{x \in W' \mid f'(x) = g'(x)\}$$

is equal to C. Then show that

$$I_{f',g'}^{W'} = I_{f,g}^W.$$

This exercise tells us that the coincidence index is completely determined by the behavior of the functions around the coincidence set. In this sense it may be viewed as a local invariant. It is of particular interest then to see how the later results will amalgamate these local invariants into a global invariant.

Now suppose $W = W_1 \cup W_2 \cup \cdots \cup W_k$ is a disjoint union of open sets and denote by C_i the compact set $C \cap W_i$ and by f_i and g_i the restrictions of f and g to W_i.

6.1 LEMMA

$$I_{f,g}^W = \sum_{i=1}^{k} I_{f_i,g_i}^{W_i}.$$

That is, the coincidence index is additive.

PROOF For each i choose an open set V_i such that $C_i \subseteq V_i \subseteq \bar{V}_i \subseteq W_i$ and set $V = \bigcup_{i=1}^{k} V_i$. Then the result follows from the commutativity of the following diagram:

$$H_n(W, W - V)$$

$$H_n(M_1) \to H_n(M_1, M_1 - V) \xrightarrow{\approx} H_n(\bigcup W_i, \bigcup (W_i - V_i)) \to H_n(M_2^\times).\quad \square$$

$$\sum_{i=1}^{k} H_n(W_i, W_i - V_i)$$

More generally, for any W let $C = C_1 \cup \cdots \cup C_k$ be a decomposition of C into disjoint compact sets. Then by repeatedly using the normality of M_1 we can find a disjoint collection of open subsets of W, denoted $W_1, \ldots, W_k$, such that $C_i \subseteq W_i$ for each i. Setting $W' = \bigcup W_i$ we can apply Lemma 6.1 and Exercise 1 to conclude that

$$I_{f,g}^W = \sum_{i=1}^{k} I_{f_i,g_i}^{W_i}.$$

6.2 LEMMA If $C = \varnothing$, then $I_{f,g}^W = 0$.

PROOF Suppose f and g have no coincidence points in the open set W. Then for any V, the map

$$(f, g): \quad (W, W - V) \to (M_2 \times M_2, M_2 \times M_2 - \Delta(M_2))$$

can be factored through the pair $(M_2 \times M_2 - \varDelta(M_2), M_2 \times M_2 - \varDelta(M_2))$ so that the induced homomorphism $(f, g)_*$ must be zero.   □ ˙

6.3   COROLLARY   If $I_{f,g}^W \neq 0$, then f and g have a coincidence in W.   □

6.4   LEMMA   Suppose f_t and $g_t \colon W \to M_2$, $0 \leq t \leq 1$, are homotopies and denote by

$$C_t = \{x \in W \mid f_t(x) = g_t(x)\}$$

for $0 \leq t \leq 1$. If $D = \bigcup_t C_t$ is a compact subset of W, then

$$I_{f_0,g_0}^W = I_{f_1,g_1}^W.$$

PROOF   Let V be an open set with $D \subseteq V \subseteq \bar{V} \subseteq W$. Then the maps

$$(f_t, g_t) \colon \quad (W, W - V) \to (M_2 \times M_2, M_2 \times M_2 - \varDelta(M_2))$$

for $0 \leq t \leq 1$ give a homotopy of maps of pairs; hence

$$(f_0, g_0)_* = (f_1, g_1)_* \quad \text{and} \quad I_{f_0,g_0}^W = I_{f_1,g_1}^W.   □$$

Exercise 2   Suppose M_1' and M_2' are closed, oriented m-manifolds and f' and $g' \colon W' \to M_2'$ are maps, where W' is open in M_1'. If

$$C' = \{y \in W' \mid f'(y) = g'(y)\}$$

is a compact subset of W', show that the coincidence index $I_{f \times f', g \times g'}^{W \times W'}$ is defined and is equal to $(I_{f,g}^W) \cdot (I_{f',g'}^{W'})$.

As a special case of this construction we may take $M_1 = M_2$ (denoted by M) and $g = $ identity on the open set W. Here a coincidence of f and g is merely a fixed point of f. For this reason the coincidence index $I_{f,\mathrm{id}}^W$ is denoted I_f^W and called the *fixed-point index* of f on W. For convenience we restate the previous results in terms of the fixed-point index.

6.5   LEMMA   If $f' \colon W' \to M$ is a map from an open set W' in M such that $f = f'$ on $W \cap W'$ and the fixed-point sets of f and f' are the same compact subset of $W \cap W'$, then $I_f^W = I_{f'}^{W'}$.   □

6.6   LEMMA   If $W = W_1 \cup W_2 \cup \cdots \cup W_k$ is a disjoint union of open subsets of M, then

$$I_f^W = \sum_{i=1}^k I_{f_i}^{W_i}.   □$$

6.7   LEMMA   If $I_f^W \neq 0$, then f has a fixed point in W. $\square$

6.8   LEMMA   If $f_t : W \to M$ is a homotopy for which the set

$$D = \{x \in W \mid f_t(x) = x \quad \text{for some} \quad 0 \le t \le 1\}$$

is compact, then $I_{f_0}^W = I_{f_1}^W$. $\square$

For most of the cases we will consider, the open set W will be the entire manifold M_1. In this case we may choose the open set V to be M_1 so that the coincidence index becomes the image of the class z_1 under the homomorphism

$$(f, g)_* : \quad H_n(M_1) \to H_n(M_2^\times) \approx Z.$$

When this occurs, the coincidence index is denoted by $I_{f,g}$ and the fixed-point index by I_f.

Example   Suppose M_1 and M_2 are closed, connected, oriented n-manifolds, $p \in M_2$, $f(M_1) = p$, and g has the property that

$$g_* : \quad H_n(M_1) \to H_n(M_2)$$

is given by $g_*(z_1) = m \cdot z_2$. We want to determine the coincidence index $I_{f,g}$.

To do this consider the following diagram:

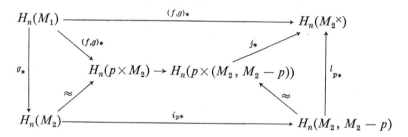

The definition of f allows us to factor $(f, g)_*$ through the upper rectangle. The commutativity of the other portions follows by using the natural identifications. Thus

$$I_{f,g} = \langle U_2, (f, g)_*(z_1) \rangle = \langle U_2, l_{p*} i_{p*} g_*(z_1) \rangle = \langle U_2, l_{p*} i_{p*} (m \cdot z_2) \rangle$$
$$= m \cdot \langle U_2, l_{p*} i_{p*}(z_2) \rangle = m \cdot \langle U_2, l_{p*}(s(p)) \rangle = m.$$

In particular, if $M_1 = M_2$ and g is the identity, then $I_f = 1$.

As another example, suppose $M_1 = M_2 = M$ and $f: W \to M$ has a single fixed point $p \in W$. We want to give another interpretation of I_f^W for this case.

First, working in euclidean space, let D^n denote the closed unit disk in $\mathbb{R}^n$. Define a map

$$F: \quad D^n \times D^n \to D^n$$

by $F(x, y) = \frac{1}{2}(y - x)$. This may be viewed geometrically in $\mathbb{R}^n \times \mathbb{R}^n$ as first taking the element of $\{0\} \times \mathbb{R}^n$ which is equivalent to (x, y) modulo the linear subspace $\Delta(\mathbb{R}^n)$ and then multiplying by $\frac{1}{2}$ (Figure 6.2).

Note that this map induces a homotopy equivalence of pairs

$$F: \quad (D^n \times D^n, D^n \times D^n - \Delta(D^n)) \to (D^n, D^n - 0).$$

To see this define $j: D^n \to D^n \times D^n$ by $j(w) = (0, w)$. The homotopy $h_t: D^n \to D^n$ given by $h_t(w) = w/(1 + t)$ has $h_0(w) = w$, $h_1(w) = F \circ j(w)$, and $h_t(D^n - 0) \subseteq D^n - 0$ for $0 \leq t \leq 1$.

On the other hand, define

$$g_t: \quad D^n \times D^n \to D^n \times D^n \quad \text{by} \quad g_t(x, y) = \left(\frac{(1 - t)x}{(1 + t)}, \frac{y - tx}{1 + t} \right).$$

It is easily checked that both coordinates lie in D^n. Then

$$g_0(x, y) = (x, y), g_1(x, y) = (0, \tfrac{1}{2}(y - x)) = j \circ F(x, y)$$

and

$$g_t(D^n \times D^n - \Delta(D^n)) \subseteq D^n \times D^n - \Delta(D^n) \quad \text{for} \quad 0 \leq t \leq 1.$$

Thus, j is a homotopy inverse for F.

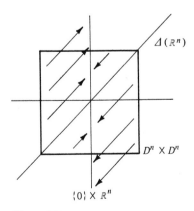

Figure 6.2

Let Y be a closed, proper n-disk in M containing p such that the homeomorphism $h: Y \to D^n$ takes p into the origin. There exists an open, proper n-disk V in M such that $p \in V$, $\bar{V} \subseteq W \cap Y$ and $f(\bar{V}) \subseteq Y$. Denote by $k: \bar{V} \to D^n$ the homeomorphism and note that k restricts to a homeomorphism of the boundaries $k: \partial \bar{V} \to S^{n-1}$. We may assume that D^n is oriented and both k and h preserve orientations.

Define a map $\phi: S^{n-1} \to S^{n-1}$ by taking the following composition:

$$S^{n-1} \xrightarrow{k^{-1}} \partial \bar{V} \xrightarrow{(f,\mathrm{id})} Y \times Y - \Delta(Y) \xrightarrow{h \times h} D^n \times D^n - \Delta(D^n) \xrightarrow{F} D^n - 0 \xrightarrow{\pi} S^{n-1},$$

where the last map is given by projecting radially from the origin.

6.9 PROPOSITION   The degree of ϕ is I_f^W.

PROOF   As we have observed, the chosen generator for $H_{n-1}(S^{n-1})$ is given by the image of the fundamental class z of M under the composition

$$H_n(M) \to H_n(M, M - V) \approx H_n(\bar{V}, \partial \bar{V}) \xrightarrow{\partial} H_{n-1}(\partial \bar{V}) \xrightarrow{k_*} H_{n-1}(S^{n-1}).$$

Note that in computing the fixed point index of f on W we may choose the open proper n-disk V. Consider the following commutative diagram:

$$H_n(M) \to H_n(M, M - V) \xrightarrow{\approx} H_n(W, W - V) \xrightarrow{(f,i)_*} H_n(M^\times) \approx Z$$

$$\approx \Big\uparrow \mathrm{incl}_* \qquad\qquad \approx \Big\uparrow \mathrm{incl}_*$$

$$H_n(\bar{V}, \partial \bar{V}) \xrightarrow{(f,i)_*} H_n(Y \times Y, Y \times Y - \Delta(Y))$$

$$\approx \Big\downarrow (h \times h)_*$$

$$H_n(D^n \times D^n, D^n \times D^n - \Delta(D^n))$$

$$\approx \Big\downarrow F_*$$

$$H_n(D^n, D^n - 0)$$

$$\approx \Big\downarrow \partial$$

$$H_{n-1}(S^{n-1}) \xrightarrow{k_*^{-1}} H_{n-1}(\partial \bar{V}) \xrightarrow{[F \circ (h \times h) \circ (f,i)]_*} H_{n-1}(D^n - 0) \xrightarrow{\approx} H_{n-1}(S^{n-1})$$

Note that since the chosen generator of $H_n(M^\times)$, $l_{p*}(s(p))$, arises from the orientation of the "vertical space" at $(p, p) \in \Delta(M)$ and the equivalence F collapses onto this vertical space, the vertical composition on the right takes $l_{p*}(s(p))$ into the chosen generator for $H_{n-1}(S^{n-1})$. Now all of the vertical homomorphisms are isomorphisms, so if the composition across the top takes z into $m \cdot l_{p*}(s(p))$, then ϕ must have degree m. $\square$

NOTE As pointed out in the introduction, this is an important step: identifying the local degree of f at an isolated fixed point (degree of ϕ) with the local fixed-point index.

Having established these basic properties of the local index, we turn our attention to the corresponding global invariants. As before, we assume that M_1 and M_2 are closed connected oriented topological n-manifolds with fundamental classes z_1 and z_2, respectively, and that f and $g: M_1 \to M_2$ are maps.

Using the coefficient homomorphism $Z \xrightarrow{\varepsilon} Q$, we denote by $\bar{z}_1$ and $\bar{z}_2$ the images of z_1 and z_2 in rational homology. That is

$$\bar{z}_1 = \varepsilon_*(z_1) \in H_n(M_1; Q) \quad \text{and} \quad \bar{z}_2 = \varepsilon_*(z_2) \in H_n(M_2; Q).$$

Consider the following diagram of groups and homomorphisms:

$$\begin{array}{ccc}
H_q(M_1; Q) & \xrightarrow{f_*} & H_q(M_2; Q) \\
\approx \Big\uparrow D_1 & & \approx \Big\uparrow D_2 \\
H^{n-q}(M_1; Q) & \xleftarrow{g^*} & H^{n-q}(M_2; Q)
\end{array}$$

where D_1 and D_2 are the Poincaré duality isomorphisms corresponding to $\bar{z}_1$ and $\bar{z}_2$, respectively. For each q define

$$\Theta_q: H_q(M_1; Q) \to H_q(M_1; Q)$$

by $\Theta_q = D_1 \circ g^* \circ D_2^{-1} \circ f_*$. Then the *Lefschetz number* or *coincidence number* of the pair (f, g) is defined to be the rational number

$$L(f, g) = \sum_q (-1)^q \operatorname{tr} \Theta_q,$$

where $\operatorname{tr} \Theta_q$ means the usual trace of Θ_q as a linear transformation from the finite-dimensional rational vector space $H_q(M_1; Q)$ to itself.

There is an alternate definition which will prove to be useful. For each q let

$$\hat{\Theta}_{n-q}: H^{n-q}(M_2; Q) \to H^{n-q}(M_2; Q)$$

be given by

$$\hat{\Theta}_{n-q} = D_2^{-1} \circ f_* \circ D_1 \circ g^*.$$

Then we define

$$\hat{L}(f, g) = \sum_{r=1}^{n} (-1)^r \operatorname{tr} \hat{\Theta}_r.$$

The relationship between these two definitions is given in the following exercises.

Exercise 3   Show that $\operatorname{tr} \Theta_q = \operatorname{tr} \hat{\Theta}_{n-q}$. Hence, conclude that

$$\hat{L}(f, g) = (-1)^n L(f, g).$$

Exercise 4   Show that $L(f, g) = (-1)^n L(g, f)$.

Recall that we have chosen a Thom class

$$U_2 \in H^n(M_2 \times M_2, M_2 \times M_2 - \Delta(M_2))$$

corresponding to the fundamental class z_2 in the manner of Chapter 5. From the composition

$$M_1 \xrightarrow{d} M_1 \times M_1 \xrightarrow{f \times g} M_2 \times M_2 \xrightarrow{i} (M_2 \times M_2, M_2 \times M_2 - \Delta(M_2))$$

we define the *Lefschetz class* or *coincidence class* of (f, g) to be

$$\mathcal{E}_{f,g} = d^* \circ (f \times g)^* \circ i^*(U_2) \in H^n(M_1).$$

Here d denotes the diagonal map. Note that the composition $(f \times g) \circ d$ is the map we have previously written as (f, g). Let $\bar{\mathcal{E}}_{f,g}$ be the image of $\mathcal{E}_{f,g}$ in rational cohomology under the coefficient homomorphism.

We now want to establish a relationship between the Lefschetz number and the Lefschetz class of (f, g). To do so we will first establish two lemmas.

Select a homogeneous basis $\{x_i\}$ for $H^*(M_2; \mathcal{Q})$ and denote by $\{a_i\}$ the basis for $H_*(M_2; \mathcal{Q})$ dual to $\{x_i\}$ under the Kronecker index. Define another basis $\{x_i'\}$ for $H^*(M_2; \mathcal{Q})$ by requiring that $D_2(x_i') = a_i$ and let $\{a_i'\}$ be the basis for $H_*(M_2; \mathcal{Q})$ dual to $\{x_i'\}$ via the Kronecker index.

Using the duality isomorphism D_1 we may define similarly related bases $\{y_i\}$ and $\{y_i'\}$ for $H^*(M_1; \mathcal{Q})$ and $\{b_i\}$ and $\{b_i'\}$ for $H_*(M_1; \mathcal{Q})$.

Suppose now that

$$f^*(x_j) = \sum_l \beta_{jl} \cdot y_l \quad \text{and} \quad g^*(x_i') = \sum_k \gamma_{ik} \cdot y_k'$$

for some rational coefficients β_{jl} and γ_{ik}.

6.10 LEMMA $\sum_i f_* \circ D_1 \circ g^*(x_i') \times a_i' = \sum_i (f \times g)_*(b_i \times b_i').$

PROOF Expanding $g_*(b_j')$ in terms of the basis $\{a_k'\}$ we have

$$g_*(b_j') = \sum_k \lambda_{kj} \cdot a_k'$$

for some rational coefficients λ_{kj}. Then note that

$$\gamma_{ij} = \left\langle \sum_k \gamma_{ik} y_k', b_j' \right\rangle = \langle g^*(x_i'), b_j' \rangle$$

$$= \langle x_i', g_*(b_j') \rangle = \left\langle x_i', \sum_k \lambda_{kj} a_k' \right\rangle = \lambda_{ij}.$$

Thus, $g_*(b_j') = \sum_i \gamma_{ij} a_i'$, or, in other words, the coefficient of y_k' in $g^*(x_i')$ is the same as the coefficient of a_i' in $g_*(b_k')$.

Using the same approach, we can show that

$$f_*(b_i) = \sum_l \beta_{li} \cdot a_l.$$

To prove the lemma, we first expand $f_* \circ D_1 \circ g^*(x_i')$ in terms of the basis $\{a_i\}$ and note that the coefficient of a_j is given by

$$\langle x_j, f_* \circ D_1 \circ g^*(x_i') \rangle = \langle f^*(x_j), D_1 \circ g^*(x_1') \rangle$$

$$= \left\langle \sum_l \beta_{jl} \cdot y_l, D_1\left(\sum_k \gamma_{ik} \cdot y_k'\right) \right\rangle$$

$$= \left\langle \sum_l \beta_{jl} \cdot y_l, \sum_k \gamma_{ik} b_k \right\rangle$$

$$= \sum_k \beta_{jk} \cdot \gamma_{ik}.$$

Thus

$$\sum_i f_* \circ D_1 \circ g^*(x_i') \times a_i' = \sum_i \left(\sum_j \left(\sum_k \beta_{jk} \cdot \gamma_{ik} \right) a_j \right) \times a_i'$$

$$= \sum_{i,j,k} (\beta_{jk} \cdot \gamma_{ik})(a_j \times a_i').$$

On the other hand,

$$\sum_i (f \times g)_*(b_i \times b_i') = \sum_i f_*(b_i) \times g_*(b_i')$$

$$= \sum_i \left(\sum_l \beta_{li} \cdot a_l \right) \times \left(\sum_k \gamma_{ki} a_k' \right)$$

$$= \sum_{i,l,k} (\beta_{li} \cdot \gamma_{ki}) \cdot (a_l \times a_k')$$

and the conclusion follows. $\square$

6.11   LEMMA   If $d: M_1 \to M_1 \times M_1$ is the diagonal and $\bar{z}_1 \in H_n(M_1; Q)$ is the fundamental class, then

$$d_*(\bar{z}_1) = \sum_i (-1)^{(\dim b_i)(\dim b_i')} \cdot b_i \times b_i'.$$

PROOF   This follows from the equations

$$(-1)^{(\dim b_j')(\dim b_k)} \cdot \langle y_k \times y_j', d_*(\bar{z}_1) \rangle$$
$$= (-1)^{(\dim y_j')(\dim y_k)} \cdot \langle y_k \cup y_j', \bar{z}_1 \rangle$$
$$= \langle y_j' \cup y_k, \bar{z}_1 \rangle$$
$$= \langle y_k, y_j' \cap \bar{z}_1 \rangle$$
$$= \langle y_k, b_j \rangle$$
$$= \delta_{k,j}. \qquad \square$$

6.12   THEOREM   The Lefschetz class $\bar{\mathfrak{S}}_{f,g}$ and the Lefschetz number $L(f, g)$ of the pair (f, g) are related by the equation

$$\langle \bar{\mathfrak{S}}_{f,g}, \bar{z}_1 \rangle = L(f, g).$$

PROOF   Since the $\{x_i'\}$ and $\{a_i'\}$ are dual bases under the Kronecker index, we may compute the Lefschetz number by

$$\hat{L}(f, g) = \sum_r (-1)^r \operatorname{tr} \hat{\Theta}_r$$
$$= \sum_i (-1)^{\dim x_i'} \langle \hat{\Theta}(x_i'), a_i' \rangle$$
$$= \sum_i (-1)^{\dim x_i'} \langle D_2^{-1} \circ f_* \circ D_1 \circ g^*(x_i'), a_i' \rangle.$$

By Equation (5.21) and the fact that $n - \dim x_i' = \dim x_i$, this is

$$\hat{L}(f, g) = \sum_i (-1)^{n(\dim x_i)(\dim x_i')} \cdot \langle i^*(\bar{U}_2), f_* \circ D_1 \circ g^*(x_i') \times a_i' \rangle$$

and by Lemma 6.10

$$\hat{L}(f, g) = \sum_i (-1)^{n(\dim b_i)(\dim b_i')} \cdot \langle i^*(\bar{U}_2), (f \times g)_*(b_i \times b_i') \rangle.$$

[The sign change, while bothersome, works out nicely, since

$$\dim f_*(b_i) = \dim b_i, \qquad \dim g_*(b_i') = \dim b_i'$$

and the sum of the two dimensions in each case is always n.]

By Lemma 6.11

$$\hat{L}(f, g) = (-1)^n \langle i^*(\bar{U}_2), (f \times g)_* d_*(\bar{z}_1) \rangle$$
$$= (-1)^n \langle d^*(f \times g)^* i^*(\bar{U}_2), \bar{z}_1 \rangle$$
$$= (-1)^n \langle \bar{\mathfrak{S}}_{f,g}, \bar{z}_1 \rangle.$$

Therefore, $L(f, g) = \langle \bar{\mathfrak{S}}_{f,g}, \bar{z}_1 \rangle.$ □

This relationship enables us to prove the following very important theorem:

6.13 LEFSCHETZ COINCIDENCE THEOREM The coincidence index of the pair (f, g) on M_1 is equal to the Lefschetz number of (f, g); that is

$$I_{f,g} = L(f, g).$$

PROOF Recall that the coincidence index is the integer given by

$$I_{f,g} = i_* \circ (f \times g)_* \circ d_*(z_1) \in H_n(M_2^\times) \approx Z,$$

where the isomorphism is given by sending a class α into the integer $\langle U_2, \alpha \rangle$.
Thus

$$I_{f,g} = \langle U_2, i_*(f \times g)_* d_*(z_1) \rangle$$
$$= \langle d^*(f \times g)^* i^*(U_2), z_1 \rangle$$
$$= \langle \mathfrak{S}_{f,g}, z_1 \rangle$$

and the image of this in the rationals is just

$$\langle \bar{\mathfrak{S}}_{f,g}, \bar{z}_1 \rangle = L(f, g)$$

by Theorem 6.12. □

Note first that this may be viewed as an "integrality" theorem. That is, the Lefschetz number $L(f, g)$ is, by definition, a rational number in general. However, its identification with the coincidence index guarantees that it will be an integer.

6.14 COROLLARY If $f, g: M_1 \to M_2$ are maps between closed oriented manifolds for which $L(f, g) \neq 0$, then f and g have a coincidence.

PROOF   This follows immediately from the fact that

$$I_{f,g} = L(f, g) \neq 0$$

and applying Corollary 6.3.   □

This is a convenient result, since in many cases the Lefschetz number is easier to compute than the coincidence index. Before proceeding with some applications, we examine a few special cases of the coincidence theorem.

First suppose $M_1 = M_2 = M$ and g is the identity map. As before the coincidence index $I_{f,\mathrm{id}}$ is written I_f and called the fixed-point index. Similarly the Lefschetz number $L(f, \mathrm{id})$ is written $L(f)$. For each k define

$$\Phi_k = f_* : \quad H_k(M; Q) \to H_k(M; Q).$$

6.15   LEMMA

$$\sum_k (-1)^k \operatorname{tr} \Phi_k = L(f).$$

PROOF   Recall that

$$L(f) = L(f, \mathrm{id}) = \sum_k (-1)^k \operatorname{tr} \Theta_k,$$

where

$$\Theta_k = D \circ \mathrm{id}^* \circ D^{-1} \circ f_* = f_*.$$

Thus, $\Phi_k = \Theta_k$ for each k, and the result follows.   □

The Lefschetz class $\mathfrak{S}_{f,\mathrm{id}}$ is written $\mathfrak{S}_f$ and, as before, its image in rational cohomology will be denoted by $\bar{\mathfrak{S}}_f$. With these definitions we have the following immediate consequences of the previous results.

6.16   LEFSCHETZ FIXED-POINT THEOREM   If $f: M \to M$ is a map of a closed, oriented n-manifold to itself, then $I_f = L(f)$. Thus, if $L(f) \neq 0$, then f has a fixed point.

PROOF   As in Theorem 6.13 we have

$$I_f = \langle \bar{\mathfrak{S}}_f, \bar{z} \rangle = L(f).$$

Then the second conclusion follows from Lemma 6.7.   □

In a further simplification we may take g and f both to be the identity on M. In this case the Lefschetz class is denoted by $\mathfrak{S}_M$ and called the

Euler class of the oriented topological manifold M. The reason for this name is readily apparent since the Lefschetz number of the identity map is the Euler characteristic, that is

$$L(\text{identity}) = \sum_k (-1)^k \operatorname{tr}(\text{id}_k)$$
$$= \sum_k (-1)^k \dim H^k(M; \mathcal{Q})$$
$$= \chi(M).$$

Thus, as a special case of the coincidence theorem (Theorem 6.13) we have established the following:

6.17 COROLLARY The value of the Euler class of M on the fundamental class of M is equal to the Euler–Poincaré characteristic of M. That is

$$\langle \mathcal{E}_M, z \rangle = \chi(M). \quad \square$$

Note that the definition of the Lefschetz number $L(f, g)$ is only dependent on the homotopy classes of the maps f and g. Thus, we can observe the following corollaries:

6.18 COROLLARY If $L(f, g) \neq 0$, g' is homotopic to g and f' is homotopic to f, then g' and f' have a coincidence. $\quad \square$

6.19 COROLLARY If $f: M \to M$ has $L(f) \neq 0$, then any map homotopic to f has a fixed point. $\quad \square$

6.20 COROLLARY If $\chi(M) \neq 0$, then any map $f: M \to M$ homotopic to the identity must have a fixed point. $\quad \square$

We now proceed with a number of applications of these theorems. First we give several fixed-point theorems due to Brouwer analogous to his theorem for the n-disk (Corollary 1.18), although slightly less well known.

6.21 COROLLARY If $f: S^n \to S^n$ is a map of degree $m \neq (-1)^{n+1}$, then f has a fixed point.

PROOF If f has degree m, then the trace of $f_*: H_n(S^n; \mathcal{Q}) \to H_n(S^n; \mathcal{Q})$ must also be m. Since the trace of $f_*: H_0(S^n; \mathcal{Q}) \to H_0(S^n; \mathcal{Q})$ is 1, we have

$$L(f) = \sum_i (-1)^i \operatorname{tr} f_*^{(i)}$$
$$= 1 + (-1)^n \cdot m.$$

Now since $m \neq (-1)^{n+1}$ we have that $L(f) \neq 0$ and f must have a fixed point by Theorem 6.16. □

Note that the antipodal map $A: S^n \to S^n$ does not have fixed points, but, as we saw in Corollary 1.22, the degree of A on S^n is $(-1)^{n+1}$.

6.22 COROLLARY If $f: RP(2n + 1) \to RP(2n + 1)$ is a map such that $f_*: H_{2n+1}(RP(2n + 1); Q) \to H_{2n+1}(RP(2n + 1); Q)$ is multiplication by $m \neq 1$, then f has a fixed point.

PROOF Note from the universal coefficient theorem that the rational homology groups of $RP(2n + 1)$ are given by

$$H_k(RP(2n + 1); Q) \approx \begin{cases} Q & \text{for} \quad k = 0, \quad 2n + 1 \\ 0 & \text{otherwise.} \end{cases}$$

Thus

$$L(f) = \sum_i (-1)^i \operatorname{tr} f_*^{(i)}$$
$$= 1 + (-1)^{2n+1} \cdot m$$
$$= 1 - m.$$

So if $m \neq 1$, $L(f) \neq 0$ and f has a fixed point. □

To see that the restriction in the theorem is necessary, consider the following function on $RP(2n + 1)$. First write S^{2n+1} in complex coordinates as $S^{2n+1} = \{(z_1, \ldots, z_{n+1}) \in C^{n+1} \mid \sum |z_j|^2 = 1\}$. Let

$$\hat{g}: \quad S^{2n+1} \to S^{2n+1}$$

be given by $\hat{g}(z_1, \ldots, z_{n+1}) = (i \cdot z_1, \ldots, i \cdot z_{n+1})$, where $i = \sqrt{-1}$. Note that $\hat{g} \circ \hat{g}$ is the antipodal map A and $\hat{g} \circ A = A \circ \hat{g}$. Thus, there is associated with $\hat{g}$ a map $g: RP(2n + 1) \to RP(2n + 1)$ for which $g^2 = $ identity.

If g has a fixed point, then there must be a nonzero $z_k = a + b \cdot i$ such that either $i \cdot z_k = z_k$ or $i \cdot z_k = -z_k$. But neither of these can happen, hence, g does not have a fixed point.

Note that $RP(2n + 1)$ is a closed orientable manifold having the same rational homology groups as a sphere of the corresponding dimension. Such manifolds are called *rational homology spheres*. It is evident that corollaries of the type of Corollaries 6.21 and 6.22 will hold for any rational homology spheres.

6.23 COROLLARY If $f: \mathbb{C}P(n) \to \mathbb{C}P(n)$ is a map for which either

(1) n is even, or
(2) $f^*: H^2(\mathbb{C}P(n)) \to H^2(\mathbb{C}P(n))$

is multiplication by $m \neq (-1)$, then f has a fixed point.

PROOF Recall that

$$H^*(\mathbb{C}P(n); \mathbb{Q}) \approx \mathbb{Q}[t]/t^{n+1},$$

where $t \in H^2(\mathbb{C}P(n); \mathbb{Q})$ is the image of an integral generator under the coefficient homomorphism. Thus, the trace of

$$f^*: \quad H^{2k}(\mathbb{C}P(n); \mathbb{Q}) \to H^{2k}(\mathbb{C}P(n); \mathbb{Q})$$

is m^k for $0 \leq k \leq n$. This implies that the trace of

$$f_*^{(2k)}: \quad H_{2k}(\mathbb{C}P(n), \mathbb{Q}) \to H_{2k}(\mathbb{C}P(n); \mathbb{Q})$$

is also m^k for $0 \leq k \leq n$. So we have

$$L(f) = \sum_i (-1)^i \, \mathrm{tr} f_*^{(i)}$$

$$= 1 + m + m^2 + \cdots + m^n$$

$$= \begin{cases} \dfrac{1 - m^{n+1}}{1 - m} & \text{if } m \neq 1 \\ n + 1 & \text{if } m = 1. \end{cases}$$

Note that if $n + 1$ is odd, this number must be nonzero. On the other hand, if $n + 1$ is even, it can only be zero when $m = -1$. Therefore, under the hypotheses of the corollary, $L(f) \neq 0$ and f has a fixed point. $\square$

Note that for $n = 1$, the antipodal map on $\mathbb{C}P(1) = S^2$ has no fixed points. Here, of course, $m = -1$.

Exercise 5 For a general odd integer n, define a map $f: \mathbb{C}P(n) \to \mathbb{C}P(n)$ that does not have fixed points.

In the same manner we may establish the following.

6.24 COROLLARY If $f: \mathbb{H}P(n) \to \mathbb{H}P(n)$ is a map for which either

(1) n is even, or
(2) $f^*: H^4(\mathbb{H}P(n)) \to H^4(\mathbb{H}P(n))$ is multiplication by $m \neq -1$,

then f has a fixed point. $\square$

Let us now investigate the situation for maps of the torus

$$T^n = \underbrace{S^1 \times S^1 \times \cdots \times S^1}_{n\text{-fold}}.$$

In this case it is necessary to change coefficients to an algebraically closed field, so let $i: Q \to C$ denote the inclusion of the rationals in the complex numbers. Using the coefficient homomorphism on homology and cohomology, the previous theorems could easily be established using complex coefficients.

Recall that $H^*(T^n; C) \approx E(C; x_1, \ldots, x_n)$, the exterior algebra over C on n generators, all of dimension 1. Thus, $f^*: H^1(T^n; C) \to H^1(T^n; C)$ is a linear transformation on an n-dimensional complex vector space. Since C is algebraically closed, there exists a basis $\{y_1, \ldots, y_n\}$ for $H^1(T^n; C)$ with respect to which the matrix $A = (a_{ij})$ of f^* is upper triangular. This basis retains the property that

$$H^*(T^n; C) \approx E(C; y_1, \ldots, y_n).$$

In $H_1(T^n; C)$ denote by $z_1, \ldots, z_n$ the dual basis. Then

$$\begin{aligned}
\langle y_k, f_*(z_i) \rangle &= \langle f^*(y_k), z_i \rangle \\
&= \langle \sum_j a_{kj} y_j, z_i \rangle \\
&= a_{ki}.
\end{aligned}$$

Thus, $f_*(z_i) = \sum_k a_{ki} z_k$, and the trace of $f_*^{(1)}$ is given by $\sum_i a_{ii}$.

Denote by $z_i \wedge z_j$ the element of $H_2(T^n; C)$ dual to the product $y_i \wedge y_j$. Then $\{z_i \wedge z_j \mid i < j\}$ is a basis for $H_2(T^n; C)$ and

$$\begin{aligned}
\langle y_i \wedge y_j, f_*(z_i \wedge z_j) \rangle &= \langle f^*(y_i \wedge y_j), z_i \wedge z_j \rangle \\
&= \langle f^*(y_i) \wedge f^*(y_j), z_i \wedge z_j \rangle \\
&= \langle (\sum_k a_{ik} y_k) \wedge (\sum_l a_{jl} y_l), z_i \wedge z_j \rangle \\
&= a_{ii} a_{jj} - a_{ij} a_{ji} = a_{ii} a_{jj}.
\end{aligned}$$

The last equality follows from the fact that $i < j$; hence, $a_{ji} = 0$. Thus, the trace of $f_*^{(2)}$ is given by

$$\sum_{i < j} a_{ii} a_{jj}.$$

Similarly we find that for $1 \leq k \leq n$ the trace of $f_*^{(k)}$ is given by

$$\sum_{i_1 < \cdots < i_k} a_{i_1 i_1} \cdots a_{i_k i_k}.$$

This implies that the Lefschetz number of f has the form

$$
\begin{aligned}
L(f) &= 1 - \sum_i a_{ii} + \sum_{i<j} a_{ii}a_{jj} - \cdots + (-1)^n a_{11}a_{22} \cdots a_{nn} \\
&= (1 - a_{11})(1 - a_{22}) \cdots (1 - a_{nn}) \\
&= \det(I - A).
\end{aligned}
$$

Therefore, we have established the following result.

6.25 COROLLARY If $f: T^n \to T^n$ is a map for which $f^*: H^1(T^n) \to H^1(T^n)$ does not have $+1$ as an eigenvalue, then f has a fixed point. $\square$

It is evident that many maps of the torus exist without fixed points. For example if $f_2: T^{n-1} \to T^{n-1}$ is any map and $f_1: S^1 \to S^1$ is a nontrivial rotation, then $f_1 \times f_2: T^n \to T^n$ has no fixed points.

Exercise 6 Let f and $g: S^n \to S^n$ be maps of degree m and k, respectively. Determine $L(f, g)$.

Exercise 7 (a) Let $f, g: S^{2n} \to \mathbb{C}P(n)$ be maps, $n > 1$. Show that $I_{f,g} = 0$.
  (b) Do there exist maps $f, g: \mathbb{C}P(n) \to S^{2n}$ such that $I_{f,g} = m$ for any integer m?

Exercise 8 Suppose that M is a closed, connected, oriented n-manifold with fundamental class $z \in H_n(M)$. If $f: M \to M$ is a map for which $f_*(z) = k \cdot z$ for some integer k, then show that

$$L(f, f) = k \cdot \chi(M).$$

The coincidence theorem gives an indirect, but appealing approach to the following basic result.

6.26 FUNDAMENTAL THEOREM OF ALGEBRA If $f(z) = z^n + a_{n-1}z^{n-1} + \cdots + a_1 z + a_0$ is a nonconstant, complex polynomial, then $f(z)$ has a root.

PROOF Denoting by $\mathbb{C}$ the complex numbers, we view f as a map from $\mathbb{C}$ to $\mathbb{C}$. Note that $|f(z)| \to \infty$ as $|z| \to \infty$; hence, we may extend f to a map of the one-point compactification

$$f: \ S^2 \to S^2$$

by setting $f(\infty) = \infty$, where ∞ denotes the north pole.

Similarly the map $g: \mathbb{C} \to \mathbb{C}$ given by $g(z) \equiv 0$ may be extended to S^2 by setting $g(\infty) = 0$, 0 being identified with the south pole. Then a coincidence of f and g will be a root of the polynomial $f(z)$.

This situation corresponds to that of the example following Lemma 6.8, so that the coincidence number $L(f, g) = k$, where

$$f_*: \ H_2(S^2) \to H_2(S^2)$$

is multiplication by k.

Certainly, if there is any justice, the degree of f should be n. To prove this, define the contracting homeomorphism $r: \mathbb{C} \to D^2 - S^1$ by $r(z) = z/(1 + |z|)$. Note that $r^{-1}(w) = w/(1 - |w|)$. There is a uniquely defined map $\hat{f}$ making the following diagram commute:

$$
\begin{array}{ccc}
\mathbb{C} & \xrightarrow{\ f\ } & \mathbb{C} \\
\downarrow{\scriptstyle r} & & \downarrow{\scriptstyle r} \\
D^2 - S^1 & \xrightarrow{\ \hat{f}\ } & D^2 - S^1
\end{array}
$$

We want to extend $\hat{f}$ to a map from D^2 to D^2. So let $w_0 \in S^1$ and let w approach w_0 through values in the interior of D^2:

$$\hat{f}(w) = r(f(r^{-1}(w)))$$

$$= r\left(f\left(\frac{w}{1 - |w|}\right)\right)$$

$$= \frac{f(w/(1 - |w|))}{1 + |f(w/(1 - |w|))|}$$

$$= \frac{w^n + a_{n-1}w^{n-1}(1 - |w|) + \cdots + a_0(1 - |w|)^n}{(1 - |w|)^n + |w^n + a_{n-1}w^{n-1}(1 - |w|) + \cdots + a_0(1 - |w|)^n|}$$

so that

$$\lim_{\substack{w \to w_0 \\ |w|<1}} \hat{f}(w) = \frac{w_0{}^n}{|w_0{}^n|} = w_0{}^n.$$

This implies that we may extend $\hat{f}$ to be defined on all of D^2 by setting $\hat{f}(w_0) = w_0{}^n$. Note that the mapping r^{-1} may be extended to a map $h: D^2 \to S^2$

by taking each point of S^1 into ∞. Then from the diagram

$$
\begin{array}{ccc}
H_2(S^2, \infty) & \xrightarrow{\ f_* \ } & H_2(S^2, \infty) \\
\approx \big\uparrow h_* & & \approx \big\uparrow h_* \\
H_2(D^2, S^1) & \xrightarrow{\ f_* \ } & H_2(D^2, S^1) \\
\approx \big\downarrow \partial & & \approx \big\downarrow \partial \\
H_1(S^1) & \xrightarrow{\ f_* \ } & H_1(S^1)
\end{array}
$$

and the fact that on S^1, $\hat{f}(e^{i\theta}) = e^{in\theta}$, we conclude that the degree of f must be n.

Therefore, $L(f, g) = n$ and f must have a root. $\quad \square$

Exercise 9 In the above setting, suppose that z_0 is a root of f of multiplicity k. Show that there exists an open set U about z_0 such that the local coincidence index $I_{f,g}^U$ is k.

The proof of Theorem 6.26 and, particularly, the accompanying exercise demonstrate that coincidence theory is a natural way to study problems of this type.

As another application of Theorem 6.16 we can prove the Poincaré–Hopf theorem on the sum of the indices of a vector field. For this purpose we must assume that our closed oriented manifold M^n is differentiable.

Let v be a smooth vector field on M^n such that the singularities (zeros) of v are isolated points of M^n. As observed before, we may associate with v a map $f: M \to M$, homotopic to the identity, having as its fixed-point set the singularities of v. If x is an isolated singularity of v, one defines the *index* of v at x, i_x, as follows. Select a coordinate neighborhood U of x,

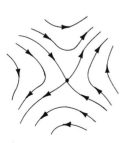

Figure 6.3

Figure 6.4

homeomorphic to an open n-disk, which contains no other singularities of v. Within U choose an $(n - 1)$-sphere about x. At each point of this sphere the associated vector of v must be nonzero. Transferring this into R^n and normalizing the vectors defines a map from S^{n-1} to S^{n-1}. The degree of this map is i_x.

For example, on a two-dimensional manifold a singularity of index -1 is shown in Figure 6.3, while the index in Figure 6.4 is $+2$. An excellent reference in this area is Milnor [1965].

It is intuitively clear that the index i_x is equal to the local degree of f at the fixed point x as defined prior to Proposition 6.9. We may use this fact to establish the following classical theorem.

6.27 POINCARÉ–HOPF THEOREM If v is a smooth vector field with isolated singularities on the closed oriented differentiable manifold M^n, then the sum of the indices of v is the Euler characteristic of M; that is

$$\sum_x i_x = \chi(M).$$

PROOF From the observations above, the sum of the local indices of v is the same as the sum of the local degrees of f at its isolated fixed points. By Proposition 6.9 this is the sum of the local fixed-point indices of f. The additivity of the fixed-point index (Lemma 6.6), together with Theorem 6.16, implies that this sum is $L(f)$. But since f is homotopic to the identity, $L(f) = \chi(M)$. □

NOTE The theorem of Poincaré mentioned in the introduction to this chapter is a special case of Theorem 6.27. Specifically, a closed surface of genus p has Euler characteristic equal to $2 - 2p$; hence, this must also be the index sum.

Having strayed this far afield, we may consider one more connection with differential topology and geometry. On a smooth manifold M we may define a cochain complex using the differential forms of M and the exterior derivative. The homology groups of the complex are the *de Rham cohomology* groups of M, denoted $H^*(M, d)$. There is a natural transformation into cohomology with real coefficients

$$\Phi: \quad H^*(M, d) \to H^*(M; \mathbb{R})$$

that may be described as follows. Suppose that M has been smoothly triangulated and ω is a smooth k-form on M. Then Φ associates with ω the function from the k-simplices of M into $\mathbb{R}$, whose value on a given simplex is the integral of ω over that simplex.

The famous *de Rham theorem* states that Φ is an isomorphism under which the exterior product in $H^*(M, d)$ corresponds to the cup product in $H^*(M; \mathbb{R})$. For a highly readable account of this, see Singer and Thorpe [1967].

Let M be a closed, connected, oriented, smooth 2-manifold endowed with a Riemannian metric. Then the *volume element* vol is a smooth 2-form on M and the *curvature K* is a smooth function associated with the Riemannian connection on M. The classical *Gauss–Bonnet theorem* then states that if the 2-form $K \cdot$ vol is integrated over the manifold M, the result is $2\pi \cdot \chi(M)$. In other words,

$$\frac{1}{2\pi} \int_M K \cdot \text{vol} = \chi(M).$$

The connection between these results and the previous is that integrating a 2-form over the manifold M corresponds under Φ with taking the Kronecker index with the fundamental class. It follows by Corollary 6.17 that the cohomology class represented by the 2-form $(1/2\pi)K \cdot$ vol is assigned by Φ to the Euler class $\mathcal{E}_M$ of M.

As a final application let M be a closed, oriented n-manifold. A *flow* on M is a one-parameter group of homeomorphisms of M. Specifically, a flow is a function

$$\phi: \quad \mathbb{R} \times M \to M$$

which is continuous and satisfies

   (i) $\phi(t_1 + t_2, x) = \phi(t_1, \phi(t_2, x))$,
   (ii) $\phi(0, x) = x$

for all $t_1, t_2 \in \mathbb{R}$ and $x \in M$.

Note that for each $t \in \mathbb{R}$ this defines a homeomorphism $\phi_t: M \to M$ by $\phi_t(x) = \phi(t, x)$ because $\phi_t^{-1} = \phi_{-t}$. A point $x_0 \in M$ is a *fixed point* of the flow if $\phi_t(x_0) = x_0$ for all $t \in \mathbb{R}$. Flows arise naturally on closed differentiable manifolds as the parameterized curves of a given vector field.

6.28   THEOREM   If M is a closed oriented manifold such that $\chi(M) \neq 0$, then any flow on M has a fixed point.

PROOF   For any $t_0 \in \mathbb{R}$ the homeomorphism

$$\phi_{t_0}: \quad M \to M$$

is homotopic (actually isotopic) to the identity. So $L(\phi_{t_0}) = L(\text{identity}) = \chi(M) \neq 0$, and ϕ_{t_0} has a fixed point.

Now for each positive integer n denote by F_n the fixed-point set of $\phi_{1/2^n}$. It follows from the additivity of the parameter that F_n will be fixed by $\phi_{m/2^n}$ for any integer m. F_n is also compact since it is the inverse image of $\varDelta(M)$ under the composition

$$M \xrightarrow{\;d\;} M \times M \xrightarrow{\;\phi_{1/2^n} \times \text{id}\;} M \times M.$$

For each positive integer n we have $F_{n+1} \subseteq F_n$ because

$$\phi_{1/2^{n+1}} \circ \phi_{1/2^{n+1}} = \phi_{1/2^n}.$$

Thus, $\{F_n\}$ is a nested family of nonempty, compact subsets of M which must have a nonempty intersection F.

This set F is fixed by ϕ_r for any dyadic rational r. Since these are dense in R, the continuity of ϕ implies that each point of F must be a fixed point of the flow ϕ.   □

Exercise 10   A flow on a manifold is the same as an *action* of the additive group of real numbers on the manifold. Using the techniques of this chapter, what results can you derive concerning actions of pathwise connected groups on closed oriented manifolds (for example, the additive group R^n or the multiplicative group S^1)?

Exercise 11   Let f and g be maps from S^3 to S^2. Show that if f is not homotopic to g, then f and g must have a coincidence.

It should be pointed out that, although the fixed-point techniques we have developed can be very useful, they are still inadequate to solve many problems. As a specific example we cite the "last geometric theorem" of Poincaré.

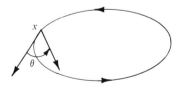

Figure 6.5

Suppose that we have an oval billiard table as in Figure 6.5, on which a single ball is rolling. Do periodic orbits with k bounces per period exist for every $k \geq 2$?

That the answer is yes was conjectured by Poincaré and proved by Birkhoff [1913].

If we orient the boundary curve, then the initial motion from the cushion is determined by the initial point x and the angle θ of projection measured from the forward pointing tangent (so $0 \leq \theta \leq \pi$). This set of initial motions given the product topology is an annulus $A = S^1 \times [0, \pi]$.

Now define

$$F: \quad A \to A$$

by taking each initial motion onto that which follows the next bounce of the orbit. The conjecture may be stated by saying that F^k has a fixed point in the interior of A for all $k \geq 2$.

In solving the problem, the following theorem was proved: Any mapping $G: A \to A$ has two fixed points in the interior of A if

(i) G is a homeomorphism leaving every point of the boundary circles fixed;

(ii) the G image of a radial 1-cell wraps at least twice around the annulus;

(iii) G preserves areas.

It is interesting to note that this problem could not have been solved by our techniques because the Lefschetz number in this case is zero.

In closing we consider briefly the question of the existence of a converse to the Lefschetz fixed-point theorem. For differentiable manifolds the Hopf theorem states that if $\chi(M) = 0$, then M admits a nonzero vector field, hence a map without fixed points which is homotopic to the identity. For topological manifolds there are a number of recent results by Brown and Fadell [1964], and others. As representatives of these results we state the following two theorems.

6.29   THEOREM   Let M be a compact, connected topological manifold. Then

(i) M admits maps close to the identity with a single fixed point;

(ii) $\chi(M) = 0$ if and only if M admits maps close to the identity without fixed points.   □

6.30 THEOREM If M is a compact, simply connected topological manifold and $f: M \to M$ is a continuous map with $L(f) = 0$, then there exists a map g homotopic to f such that g is fixed-point free. $\quad\square$

For a deeper, more comprehensive study of fixed-point theory see Brown [to appear].

appendix I

The purpose of this appendix is to give a proof of Theorem 1.14. The proof requires the development of the subdivision operators on the chain groups. This fundamental technique is at the basis of essentially all of the computations and applications we will be able to make.

If $C \subseteq R^n$ is a bounded set, then the *diameter* of C is given by diam $C = \mathrm{lub}\{\| x - y \| \mid x, y \in C\}$. If $\mathcal{C} = \{C_i\}$ is a family of bounded subsets of R^n, then mesh $\mathcal{C} = \mathrm{lub}\{\mathrm{diam}\, C_i\}$.

I.1 PROPOSITION If s^n is an n-simplex with vertices $a_0, a_1, \ldots, a_n$, then diam $s^n = \max\{\| a_i - a_j \| \mid i, j = 0, \ldots, n\}$.

PROOF Let $x = \sum t_i a_i$ and $y = \sum t_i' a_i$ be points in s^n. First fix x and allow y to vary. We want to show that

$$\mathrm{lub}_{y \in s^n} \| x - y \| = \mathrm{lub}_i \| x - a_i \|.$$

Now

$$\| x - y \| = \| x - \sum t_i' a_i \| = \| \sum t_i'(x - a_i) \|$$
$$\leq \sum | t_i' | \cdot \| x - a_i \| = \sum t_i' \| x - a_i \|$$
$$\leq \sum t_i' \cdot \max\{\| x - a_i \|\} = \max\| x - a_i \|.$$

Repeating the above, letting x vary gives

$$\| x - y \| \leq \max \| a_j - a_i \|. \quad \square$$

Let s^n be an n-simplex with vertices $a_0, a_1, \ldots, a_n$. The *barycenter* $b(s^n)$ of s^n is the point in s^n given by

$$b(s^n) = (1/(n + 1))(a_0 + \cdots + a_n).$$

It is not difficult to show that for any i with $0 \leq i \leq n$, the points

$$\{b(s^n), \ a_0, \ \ldots, \ a_{i-1}, \ a_{i+1}, \ \ldots, \ a_n\}$$

span an n-simplex. We now define the *barycentric subdivision* $\mathrm{Sd}(s^n)$ inductively on the dimension of the simplex. First set $\mathrm{Sd}(s^0) = s^0$ for any zero simplex s^0. Suppose now that Sd is defined on any simplex of dimension $(n - 1)$, so if t^{n-1} is any $(n - 1)$-simplex, $\mathrm{Sd}(t^{n-1})$ is a collection of $(n - 1)$-simplices geometrically contained in t^{n-1}. Denote by $\dot{s}^n$ the collection of all $(n - 1)$-faces of s^n and define

$$\mathrm{Sd}(\dot{s}^n) = \bigcup_{t^{n-1} \in \dot{s}^n} \mathrm{Sd}(t^{n-1}).$$

Then $\mathrm{Sd}(s^n)$ will consist of all n-simplices of the form $(b(s^n), t_0, \ldots, t_{n-1})$, where $(t_0, \ldots, t_{n-1})$ is an $(n - 1)$-simplex in $\mathrm{Sd}(\dot{s}^n)$ (Figure I.1).

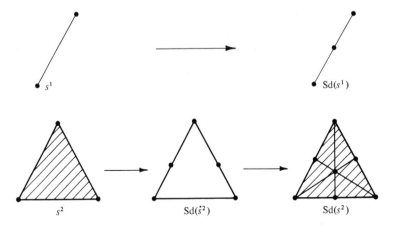

Figure I.1

I.2   PROPOSITION If K is a collection of n-simplices, then

$$\text{mesh Sd}(K) \leq (n/(n+1)) \text{ mesh } K.$$

PROOF Proceeding by induction on n, for $n = 0$ both sides are zero. So suppose the result is true for any collection of $(n-1)$-simplices. Let t^n be an n-simplex in $\text{Sd}(K)$. Then $t^n = (b(s^n), u_0, \ldots, u_{n-1})$ where $s^n \in K$ and $u_0, \ldots, u_{n-1}$ are vertices of an $(n-1)$-simplex ω in $\text{Sd}(\dot{s}^n)$. Let s^{n-1} be the $(n-1)$-simplex in $\dot{s}^n$ containing ω.

By Proposition I.1

$$\text{diam } t^n = \max\{\| u_i - u_j \|, \| u_i - b(s^n) \|\}.$$

First consider the terms of the form $\| u_i - u_j \|$. We know that

$$\| u_i - u_j \| \leq \text{diam } \omega \leq ((n-1)/n) \text{ diam } s^{n-1}$$

by the inductive hypothesis. Now since $x/(x+1)$ is an increasing function and the diameter of a subset is less than or equal to the diameter of the set, we have

$$\frac{n-1}{n} \text{ diam } s^{n-1} \leq \frac{n}{n+1} \text{ diam } s^n.$$

Hence, any term of the form $\| u_i - u_j \|$ is less than or equal to $(n/(n+1))$ diam s^n.

For the terms $\| u_i - b(s^n) \|$ recall that if $u_0', \ldots, u_n'$ are the vertices of s^n then $b(s^n) = (1/(n+1)) \cdot \sum u_i'$. Each vertex u_i is a point in s^n so that $\| u_i - b(s^n) \| \leq \| u_j' - b(s^n) \|$ for some j by the proof of Proposition I.1. Then

$$\| u_j' - b(s^n) \| = \left\| u_j' - \frac{1}{n+1} \sum_i u_i' \right\| = \left\| \sum_{i \neq j} \frac{u_j' - u_i'}{n+1} \right\|,$$

where the sum now has n terms. But then

$$\left\| \sum_{i \neq j} \frac{u_i' - u_j'}{n+1} \right\| \leq \frac{1}{n+1} \sum_{i \neq j} \| u_j' - u_i' \|$$

$$\leq \frac{n}{n+1} \max \| u_j' - u_i' \|$$

$$\leq \frac{n}{n+1} \text{ diam } s^n.$$

Thus, all terms will satisfy the desired inequality and the proof is complete. $\square$

I.3 COROLLARY If K is a collection of n-simplices, let $\mathrm{Sd}^m(K) = \mathrm{Sd}(\mathrm{Sd}^{m-1}(K))$ be the iterated barycentric subdivision. Then for an n-simplex s^n and any $\varepsilon > 0$ there exists a positive integer m such that

$$\text{mesh } \mathrm{Sd}^m(s^n) < \varepsilon.$$

PROOF This follows immediately from Proposition I.2 and the fact that

$$\lim_{m \to \infty} \left(\frac{n}{n+1} \right)^m = 0. \qquad \square$$

With these basic properties of the subdivision operator on simplices in mind, we now want to define an analogous operation on singular simplices. If C and C' are convex sets, a map $f: C \to C'$ is *affine* if given $x, y \in C$, $0 \leq t \leq 1$, then

$$f((1 - t)x + ty) = (1 - t)f(x) + tf(y).$$

It follows from this that if $x_0, \ldots, x_p \in C$ and $t_0, \ldots, t_p$ are nonnegative with $\sum t_i = 1$, then

$$f(\sum t_i x_i) = \sum t_i f(x_i).$$

If C is convex, define $A_n(C) \subseteq S_n(C)$ to be the subgroup generated by all affine singular n-simplices $\phi: \sigma_n \to C$. Denoting by $v_0, v_1, \ldots, v_n$ the vertices of σ_n, for any affine $\phi: \sigma_n \to C$ let $x_i = \phi(v_i)$. Then we can denote ϕ by $x_0 x_1 \cdots x_n$. In this notation it is evident that

$$\partial_i(x_0 x_1 \cdots x_n) = (x_0 \cdots x_{i-1} x_{i+1} \cdots x_n).$$

Thus, $\partial(A_n(C)) \subseteq A_{n-1}(C)$ and $\{A_n(C)\} = A_*(C)$ is a chain complex.

We now define a chain map $\mathscr{Sd}': A_n(C) \to A_n(C)$ which is the algebraic analog of the subdivision operation. The definition is given inductively on the dimension n. For $n = 0$ let $\mathscr{Sd}'$ be the identity. Suppose now that it is defined up through dimension $n - 1$, and let $\phi = x_0 x_1 \cdots x_n$ be an affine singular n-simplex in C. The *barycenter* of ϕ is the point

$$b(\phi) = \frac{x_0 + \cdots + x_n}{n + 1}.$$

For any point $b \in C$ define a homomorphism

$$\mathscr{C}_b: \quad A_{n-1}(C) \to A_n(C)$$

by

$$\mathcal{C}_b(y_0y_1 \cdots y_{n-1}) = (by_0y_1 \cdots y_{n-1}).$$

This is called the *cone* on b for obvious geometrical reasons. Finally, define for any affine singular n-simplex ϕ (see Figure I.2)

$$\mathscr{S}\!d'(\phi) = \mathcal{C}_{b(\phi)}(\mathscr{S}\!d'(\partial\phi)).$$

I.4 PROPOSITION $\partial \circ \mathscr{S}\!d' = \mathscr{S}\!d' \circ \partial.$

PROOF It is sufficient to check this on some affine singular n-simplex $\phi = x_0x_1 \cdots x_n$. Let $b = b(\phi)$. Certainly the formula is true in dimension $n = 0$, so assume that it holds in dimension $(n - 1)$,

$$\partial \mathscr{S}\!d'(x_0 \cdots x_n) = \partial\mathcal{C}_b(\mathscr{S}\!d'\partial(x_0 \cdots x_n)).$$

We may split up the boundary on the right into those terms containing b and those not containing b,

$$\partial \mathscr{S}\!d'(x_0 \cdots x_n) = \mathscr{S}\!d'\partial(x_0 \cdots x_n) - \mathcal{C}_b(\partial \mathscr{S}\!d'\partial(x_0 \cdots x_n)).$$

But the second term here must be zero, for by the inductive hypothesis

$$\partial \mathscr{S}\!d'\partial(x_0 \cdots x_n) = \mathscr{S}\!d'\partial\partial(x_0 \cdots x_n) = 0.$$

Therefore

$$\partial \mathscr{S}\!d' = \mathscr{S}\!d'\partial. \qquad \square$$

Thus, $\mathscr{S}\!d': A_*(C) \to A_*(C)$ is a chain map of degree zero. Since the homology should not be affected by subdividing simplices, it is reasonable to expect that $\mathscr{S}\!d'$ is chain homotopic to the identity. To verify that this is indeed the case we must define a homomorphism

$$T': \quad A_n(C) \to A_{n+1}(C)$$

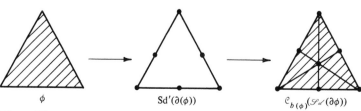

$$\phi \qquad\qquad \mathrm{Sd}'(\partial(\phi)) \qquad\qquad \mathcal{C}_{b(\phi)}(\mathscr{S}\!d'(\partial\phi))$$

Figure I.2

such that

$$\partial T' + T'\partial = \mathcal{S}\!\mathpzc{d}' - 1.$$

We define T' inductively on n. Since for $n = 0$, $\mathcal{S}\!\mathpzc{d}'$ is the identity, we take T' to be zero. Now suppose T' is satisfactorily defined on all chains of dimension less than n and let ϕ be an affine singular n-simplex.

Note that

$$\partial(\mathcal{S}\!\mathpzc{d}'\phi - \phi - T'\partial\phi) = [\partial\mathcal{S}\!\mathpzc{d}' - \partial - (\mathcal{S}\!\mathpzc{d}' - 1 - T'\partial)\partial]\phi$$

$$= 0 \quad \text{since } \mathcal{S}\!\mathpzc{d}'\partial = \partial\mathcal{S}\!\mathpzc{d}' \text{ and } \partial\partial = 0.$$

Then set

$$T'(\phi) = \mathcal{C}_{b(\phi)}(\mathcal{S}\!\mathpzc{d}'\phi - \phi - T'\partial\phi).$$

To compute $\partial T'(\phi)$ we split it up into that part containing $b(\phi)$ and that part not containing $b(\phi)$. In other words,

$$\partial T'(\phi) = (\mathcal{S}\!\mathpzc{d}'\phi - \phi - T'\partial\phi) - \mathcal{C}_{b(\phi)}\partial(\mathcal{S}\!\mathpzc{d}'\phi - \phi - T'\partial\phi).$$

But from the above computation, the second term must be zero and T has the desired property.

If $f\colon C \to C'$ is an affine map between convex sets, then

$$f_{\#}(A_n(C)) \subseteq A_n(C')$$

and $f_{\#}$ commutes with both $\mathcal{S}\!\mathpzc{d}'$ and T'.

We want to use the above homomorphisms to construct a degree-zero chain map

$$\mathcal{S}\!\mathpzc{d}\colon\ S_n(X) \to S_n(X)$$

for any space X, and show that it is chain homotopic to the identity.

Using the same technique as in the proof of Theorem 1.10, let $\psi\colon \sigma_n \to X$ be a singular n-simplex. There is the induced homomorphism

$$\psi_{\#}\colon\ S_n(\sigma_n) \to S_n(X).$$

Now the element $\tau_n \in S_n(\sigma_n)$ given by the identity map is in $A_n(\sigma_n)$. So define

$$\mathcal{S}\!\mathpzc{d}(\psi) = \mathcal{S}\!\mathpzc{d}\ \psi_{\#}(\tau_n) = \psi_{\#}\mathcal{S}\!\mathpzc{d}'(\tau_n).$$

This just has the effect of subdividing the simplex by subdividing in its domain, which is convex. Similarly, set

$$T(\psi) = T\psi_{\#}(\tau_n) = \psi_{\#}T'(\tau_n).$$

I.5 PROPOSITION $\partial T + T\partial = \mathscr{S}d - 1.$

PROOF This follows immediately from the same properties of T' and $\mathscr{S}d'$. $\square$

We are now ready to give a proof of the theorem.

1.14 THEOREM If $\mathfrak{U}$ is a family of subsets of X such that Int $\mathfrak{U}$ covers X, then the chain map $i: S_*(X) \to S_*(X)$ induces an isomorphism

$$i_*: \quad H_n(S_*^{\mathfrak{U}}(X)) \to H_n(X).$$

PROOF We will construct a chain map $\Phi: S_*(X) \to S_*^{\mathfrak{U}}(X)$ such that $\Phi \cdot i$ is the identity and $i \cdot \Phi$ is chain homotopic to the identity.

Let $\phi: \sigma_n \to X$ be a singular n-simplex. The family $\mathfrak{V} = \{\phi^{-1}(U)|\ U \in \mathfrak{U}\}$ has Int $\mathfrak{V}$ covering σ_n. Since σ_n is compact, there exists a $\delta > 0$ such that if $C \subseteq \sigma_n$ and diam $C < \delta$, then C is contained in $\phi^{-1}(U)$ for some U. By Corollary I.3 there exists an $m \geq 0$ with

$$\text{mesh Sd}^m\ \sigma_n < \delta.$$

This will imply that

$$\mathscr{S}d^m\ \phi \in S_n^{\mathfrak{U}}(X).$$

Now for any singular simplex ϕ in X let $m(\phi)$ be the least integer for which

$$\mathscr{S}d^{m(\phi)}\phi \in S_n^{\mathfrak{U}}(X).$$

Note that for $0 \leq i \leq n$, $m(\phi) \geq m(\partial_i\phi)$.

Recall that $\partial T + T\partial = \mathscr{S}d - 1$, so for any positive integer k we have

$$\partial T\mathscr{S}d^{k-1} + T\mathscr{S}d^{k-1}\partial = \mathscr{S}d^k - \mathscr{S}d^{k-1}.$$

Adding a sequence of these together gives

$$\partial T(1 + \cdots + \mathscr{S}d^{k-1}) + T(1 + \cdots + \mathscr{S}d^{k-1})\partial = \mathscr{S}d^k - 1.$$

So we define for any ϕ

$$\mathfrak{I}(\phi) = T(1 + \mathscr{S}d + \cdots + \mathscr{S}d^{m(\phi)-1}),$$

and consider

$$(\partial\mathfrak{I} + \mathfrak{I}\partial)\phi = \sum (-1)^i\partial_i T(1 + \cdots + \mathscr{S}d^{m(\phi)-1})\phi$$
$$+ \sum (-1)^i T(1 + \cdots + \mathscr{S}d^{m(\partial_i\phi)-1})\partial_i\phi.$$

By the above we have

$$(\partial \mathfrak{I} + \mathfrak{I}\partial)\phi = \mathscr{S}\mathscr{A}^{m(\phi)}\phi - \phi - T(1 + \cdots + \mathscr{S}\mathscr{A}^{m(\phi)-1}) \, \partial\phi$$
$$+ \sum (-1)^i T(1 + \cdots + \mathscr{S}\mathscr{A}^{m(\partial_i\phi)-1}) \, \partial_i\phi$$
$$= \mathscr{S}\mathscr{A}^{m(\phi)}\phi - \phi - \sum_{i=0}^{n} (-1)^i T(\mathscr{S}\mathscr{A}^{m(\partial_i\phi)} + \cdots + \mathscr{S}\mathscr{A}^{m(\phi)-1})\partial_i\phi.$$

This leads us to define

$$\Phi(\phi) = \mathscr{S}\mathscr{A}^{m(\phi)}\phi - \sum_{i=0}^{n} (-1)^i T(\mathscr{S}\mathscr{A}^{m(\partial_i\phi)} + \cdots + \mathscr{S}\mathscr{A}^{m(\phi)-1})\partial_i\phi.$$

From looking at the summation we conclude that

$$\Phi(\phi) \in S_n^{\,\mathfrak{U}}(X).$$

To consider $\Phi(\phi)$ as an element of $S_n(X)$ we apply the mapping i. The above manipulation shows that

$$\partial \mathfrak{I} + \mathfrak{I}\partial = i \circ \Phi - 1;$$

hence, $i \circ \Phi$ is chain homotopic to the identity. On the other hand, if $\phi \in S_n^{\,\mathfrak{U}}(X)$, then $m(\phi) = 0$ and $\Phi \circ i$ is the identity. $\square$

appendix II

The purpose of this appendix is to prove two of the basic theorems on topological manifolds which were used in Chapter 5. The first theorem states that any closed topological manifold may be imbedded in some euclidean space R^n and that the imbedded manifold is a retract of a neighborhood in R^n. The second theorem states that the boundary of a compact topological manifold admits a collaring.

The first result is an excellent example of a "folk" theorem. That is, a result which is well known and may be proved in a variety of ways, but which is difficult to locate in the literature or trace to its true origin. The imbedding technique we use is due to Dold, whereas the approach to the retraction property was suggested by Bing.

The second result is of more recent vintage. A collaring theorem for differentiable manifolds was proved by Milnor, and the analog for topological manifolds, by Brown [1962]. The proof we present here is a recent one due to Conelly [1971].

II.1 THEOREM If M is a closed topological n-manifold, then M can be imbedded in euclidean space R^k for some k.

PROOF Let $B_1, \ldots, B_m$ be a collection of proper open n-balls in M which cover M. For $i = 1, \ldots, m$ denote by

$$\bar{h}_i: \quad B_i \to S^n - \{y\}$$

a homeomorphism onto the complement of the north pole. We can extend each $\bar{h}_i$ to a map

$$h_i: \quad M \to S^n$$

by defining

$$h_i(x) = \begin{cases} \bar{h}_k(x) & \text{if} \quad x \in B_i \\ \{y\} & \text{if} \quad x \in M - B_i. \end{cases}$$

Now define the map

$$i: \quad M \to S^n \times S^n \times \cdots \times S^n \subseteq \mathbb{R}^{n+1} \times \mathbb{R}^{n+1} \times \cdots \times \mathbb{R}^{n+1} = \mathbb{R}^{m(n+1)}$$

by $i(x) = (h_1(x), h_2(x), \ldots, h_m(x))$. Then i gives the desired imbedding. $\square$

NOTE In general this is not a very economical way to imbed the manifold. That is, the dimension of the euclidean space is much higher than is generally necessary. For example, the covering of a circle by two proper 1-balls will produce the imbedding illustrated in Figure II.1 (actually in $\mathbb{R}^4$).

Suppose now that $i: M^n \to \mathbb{R}^k$ is an imbedding of a closed topological n-manifold. Denote by s a large k-simplex in $\mathbb{R}^k$ containing M in its interior. We want to triangulate the complement of M in s in a particular way. Denote by Sd the barycentric subdivision operator defined in Appendix I and let $s_1 = \mathrm{Sd}(s)$, a simplicial complex that is a finite union of k-simplices.

Now examine each closed k-simplex in s_1. To those which intersect M we apply the operator Sd. Those which do not intersect M are left intact. The resulting simplicial complex is denoted s_2.

By continuing this process, we produce a sequence of finite simplicial

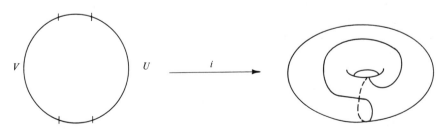

Figure II.1

complexes $\{s_n\}$, each a finite union of closed k-simplices, with the property that s_m subdivides s_n whenever $m \geq n$ (Figure II.2).

II.2 LEMMA This process defines a triangulation of $s - M$. In other words, for every point x of $s - M$ there is an integer m such that each k-simplex containing x in s_m remains intact in $s_{m'}$ for all $m' \geq m$.

PROOF Let $x \in s - M$. Since M is compact, the distance from x to M is some positive number ε. By Corollary I.3 there is a positive integer m such that

$$\text{mesh Sd}^m(s) < \varepsilon.$$

There are two possibilities:

(a) at some stage s_l, $l < m$, all of the closed k-simplices containing x were disjoint from M; or

(b) at each stage s_l, $l < m$, some closed k-simplex containing x intersected M.

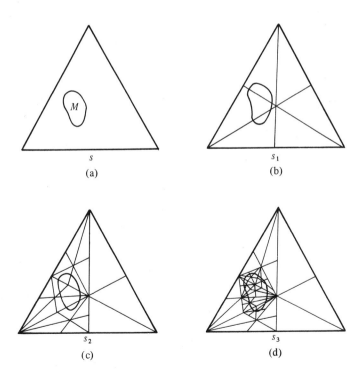

Figure II.2

In the first case, for all stages beyond s_l the triangulation around x remains unchanged. In the second case, each k-simplex of s_{m-1} that intersects M will be a k-simplex of $\mathrm{Sd}^{m-1}(s)$. Thus, the k-simplices of s_m containing x will either be k-simplices of s_{m-1} that do not intersect M, or barycentric subdivisions of those that did, hence k-simplices of $\mathrm{Sd}^m(s)$. In either situation it follows that the k-simplices of s_m that contain x will be disjoint from M.

Therefore, in each s_j, $j \geq m$, the triangulation about x remains constant. In this way we define the triangulation of $s - M$. □

We now want to use this triangulation to define inductively a collection of subsets $N_0 \subseteq N_1 \subseteq \cdots \subseteq N_k = N$ of s together with a map $r \colon N \to M$. Initially we take N_0 to be the union of M with all the vertices of $s - M$. If $y \in M$, define $r(y) = y$. If x_0 is a vertex of $s - M$, define $r(x_0)$ to be some point of M for which $\mathrm{dist}(x_0, r(x_0)) = \mathrm{dist}(x_0, M)$.

Suppose then that N_{i-1} and $r \colon N_{i-1} \to M$ have been defined. Let α be a closed i-simplex in $s - M$; α will be contained in N_i if both of the following requirements are satisfied:

(a)   the boundary of α is contained in N_{i-1};

(b)   the map $r \colon \partial\alpha \to M$ can be extended continuously over α.

The space N_i will be the union of N_{i-1} together with all such closed i-simplices α. To define $r \colon N_i \to M$ it is sufficient to define r on each α so that it is compatible with the previous definition on the boundary. Let

$$A = \{\delta \in \mathbb{R} \mid \text{there is a map } f \colon \alpha \to M, f|_{\partial\alpha} = r, \text{ and } \mathrm{diam}(\text{image } f) = \delta\}$$

For α an i-simplex in N_i, A is a nonempty set that is bounded below. Let a be the greatest lower bound for A. Now define $r \colon \alpha \to M$ by choosing a map that extends the restriction of r to $\partial\alpha$ and satisfies $a \leq \mathrm{diam}(r(\alpha)) \leq 2a$. Note that if $a = 0$, we may take r to be the constant map taking α into $r(\partial\alpha)$.

This completes the inductive step so that $N = N_k$ is a well-defined subset of s containing M and $r \colon N \to M$ is a function which is the identity on M. It remains to be shown that r is continuous and N is a neighborhood of M in $\mathbb{R}^k$. First we present some preliminary lemmas.

II.3   LEMMA   For any $\varepsilon > 0$ there exists a $\delta > 0$ such that if x is a point of $s - M$ and $\mathrm{dist}(x, M) < \delta$, then the mesh of the set of k-simplices of $s - M$ containing x is less than ε.

PROOF   Let $\varepsilon > 0$; choose a positive integer m so that $(k/(k + 1))^m \cdot$ diam$(s) < \varepsilon$. Define K_m to be the union of all closed k-simplices of s_m that do not intersect M. K_m is a compact subset of s.

Let $\delta = \text{dist}(M, K_m)$, or, if K_m is empty, take $\delta = 1$. Then if dist$(x, M) < \delta$, each closed k-simplex of s_m that contains x must intersect M. Thus, each of these simplices must lie in $\text{Sd}^m(s)$ and their mesh is less than or equal to mesh $\text{Sd}^m(s) \leq (k/(k + 1))^m \cdot$ diam $s < \varepsilon$. The same inequality will obviously be true for the set of k-simplices of $s - M$ containing x.   □

II.4   LEMMA   If $\mathcal{S}$ is a p-simplex, K is a convex set, and $f: \partial\mathcal{S} \to K$ is a map, then f can be extended over all of $\mathcal{S}$.

PROOF   Let $b(\mathcal{S})$ be the barycenter of $\mathcal{S}$ and select a point $w \in K$. Every point x of $\mathcal{S}$ has a unique representation in the form $x = ty + (1 - t) \cdot b(\mathcal{S})$, where y is a point of $\partial\mathcal{S}$ and $0 \leq t \leq 1$. For each such point x define

$$f(x) = t \cdot f(y) + (1 - t)w.$$

This is well defined since K is convex; f is continuous and extends the original definition of f on $\partial\mathcal{S}$.   □

II.5   COROLLARY   If $\mathcal{S}$ is a p-simplex and B is a proper n-ball in a topological manifold, then any map $f: \partial\mathcal{S} \to B$ can be extended over all of $\mathcal{S}$.

PROOF   Since B is a proper n-ball, there is a homeomorphism

$$h: \quad B \to D^n - S^{n-1},$$

a convex subset of $\mathbb{R}^n$. We may compose f with h and apply Lemma II.4 to give the desired extension.   □

II.6   THEOREM   The function $r: N \to M$ defined previously is a continuous retraction of a neighborhood of M in $\mathbb{R}^k$ onto M.

PROOF   From the construction of r it is apparent that r is continuous at each point of $s - M$. Thus, it is sufficient to check continuity at a point $y \in M$. So let $\varepsilon > 0$ and denote by $B(y, \varepsilon)$ the ball of radius ε about $y = r(y)$ in $\mathbb{R}^k$.

We construct inductively a collection of open sets $V_0, V_1, \ldots, V_{k+1}$ in $\mathbb{R}^k$ about y. Let $V_0 = B(y, \varepsilon)$. For the inductive step, suppose that V_{l-1}

has been defined. Let V_l be an open subset of V_{l-1} in R^k containing y having the following properties:

  (i) $V_l \subseteq B(y, \delta_l)$, where $B(y, 5\delta_l) \subseteq V_{l-1}$;
  (ii) $V_l \cap M$ is a proper n-ball about y;
  (iii) if $x \in V_l - M$, then the mesh of the set of k-simplices of $s - M$ containing x is $\leq \delta_l$.

That Requirement (iii) may be satisfied follows from Lemma II.3.

Now let $x \in V_{k+1}$, an open set about y. If $x \in M$, then

$$r(x) = x \in V_0 = B(y, \varepsilon).$$

So suppose $x \notin M$ and let $\mathfrak{S}$ be a k-simplex of $s - M$ containing x.

By Requirements (iii) and (i) all of the vertices of $\mathfrak{S}$ must lie in V_k and must be mapped by r into the proper n-ball $V_k \cap M$. By Corollary II.5 each 1-simplex α in $\mathfrak{S}$ admits a map into $V_k \cap M$ extending the restriction of r to $\partial\alpha$. Thus, each set α is contained in N. Furthermore, since the diameter of the image of this extension is less than $2\delta_k$, the diameter of $r(\alpha)$ must be less than $4\delta_k$. The fact that $r(\alpha)$ intersects $V_k \cap M$, together with Requirement (i), implies that $r(\alpha) \subseteq V_{k-1} \cap M$.

Thus, the image of the 1-skeleton of $\mathfrak{S}$ under r is contained in $V_{k-1} \cap M$, a proper n-ball. We may now apply the same argument to the 2-simplices of $\mathfrak{S}$. Continuing inductively, we find that the image of the k-skeleton of $\mathfrak{S}$ under r, that is, $r(\mathfrak{S})$, is contained in $V_{k-k} = V_0 = B(y, \varepsilon)$. In particular $r(x) \in B(y, \varepsilon)$ and r is continuous at y.

To see that N is a neighborhood of M in R^k, note that in the above argument, V_{k+1} is an open set about $y \in M$ on which r is completely defined. Hence, $V_{k+1} \subseteq N$ and y is an interior point of N. $\square$

Exercise 1   Make the necessary modifications in the preceding proofs to show that the results hold as well for compact manifolds with boundary.

Exercise 2   Prove a similar imbedding and retraction theorem for non-compact manifolds.

Finally, we turn to the collaring theorem for topological manifolds. As we stated previously, the proof given here is an intuitively appealing one due to Conelly [1971].

II.7   THEOREM (*Topological collaring theorem*)   Let M^n be a compact topological manifold with boundary $\partial M = B$. Then there exists an open set U in M, containing B, and a homeomorphism

$$h: \quad U \to B \times [0, 1)$$

such that $h(x) = (x, 0)$ for all $x \in B$.

PROOF   The idea of the proof is as follows. Since $B = \partial M$ we can find about each point of B an open set which looks like a portion of a "collar"; that is, B is locally collared in M. We attach a collar to the boundary of M and then use the local collaring to push the added collar into the manifold (or pull the manifold out over it), so that the added collar becomes the desired open set U.

By using the topological properties of euclidean half-space H^n, we can show that for any point $x \in B = \partial M$ there is an open set U_x in B about x and an imbedding

$$h_x: \quad \bar{U}_x \times [0, 1] \to M$$

such that for any $x' \in \bar{U}_x$, $h_x(x', 0) = x'$. Now B is a closed subspace of the compact manifold M; hence, B is compact and there exist a finite number of open sets $U_1, \ldots, U_m$ in B and imbeddings

$$\bar{h}_i: \quad \bar{U}_i \times [0, 1] \to M$$

such that

$$\bar{h}_i(x, 0) = x \qquad \text{for all} \quad x \quad \text{in} \quad \bar{U}_i$$

and

$$B = \bigcup_i U_i.$$

Since B is compact Hausdorff, it is also normal; hence, there exist open sets $V_1, \ldots, V_m$ covering B such that $\bar{V}_i \subseteq U_i$ for each i.

Define M^+ to be the space formed from the union $M \cup (B \times [-1, 0])$ by identifying $x \in B \subseteq M$ with $(x, 0) \in B \times [-1, 0]$ (Figure II.3). For

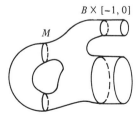

Figure II.3

each i let

$$h_i: \quad \bar{U}_i \times [-1, 1] \to M^+$$

be the function given by

$$h_i(x, t) = \begin{cases} \bar{h}_i(x, t) & \text{if } 0 \le t \le 1 \\ (x, t) & \text{if } -1 \le t \le 0. \end{cases}$$

Since these agree on the intersection, each h_i is a well-defined imbedding.

We now use these maps h_i to define inductively a family of imbeddings $g_i: M \to M^+$ and maps $f_i: B \to [-1, 0]$, $i = 0, 1, 2, \ldots, m$, satisfying the following:

(a) $g_i(M)$ contains $M \cup (\bigcup_{j \le i} \bar{V}_j \times [-1, 0])$;

(b) for any $x \in B$, $g_i(x) = (x, f_i(x))$;

(c) $f_m(x) \equiv -1$;

(d) for any $x \in B$, $\{x\} \times [f_i(x), 0] \subseteq g_i(M)$.

The imbeddings g_i correspond to the consecutive stages of pushing the collar into the manifold while the functions f_i keep track of the location of the boundary of M at each stage. It follows that g_m will be a homeomorphism of M with M^+ taking $x \in B$ to $(x, -1)$ in M^+. This will give the desired collaring of B.

Define $g_0: M \to M^+$ to be the inclusion and set $f_0(B) \equiv 0$. Inductively, suppose g_{i-1} and f_{i-1} have been defined. Consider

$$h_i^{-1}(g_{i-1}(M)) \subseteq \bar{U}_i \times [-1, 1]$$

(for example, the shaded region in Figure II.4). We want to define an

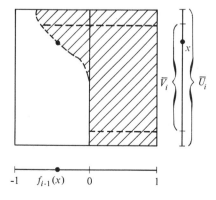

-1 $f_{i-1}(x)$ 0   1

Figure II.4

imbedding

$$\phi_i\colon\ h_i^{-1}(g_{i-1}(M)) \to \bar{U}_i \times [-1, 1]$$

by pushing to the left along the fibers until $\phi_i(h_i^{-1}(g_{i-1}(\bar{V}_i))) = \bar{V}_i \times \{-1\}$, but requiring that ϕ_i be the identity on $(\bar{U}_i - U_i) \times [-1, 1] \cup \bar{U}_i \times \{1\}$. Thus, ϕ_i represents a "pushing out" operation inside this local collar which will not effect the rest of the manifold.

To do this, we want a map $\lambda_i\colon \bar{U}_i \to [-1, 1]$ such that

$$\lambda_i(x) = \begin{cases} 2f_{i-1}(x) + 1 & \text{if } x \in \bar{V}_i \\ -1 & \text{if } x \in \bar{U}_i - U_i \end{cases}$$

and $\lambda_i(x) \leq 2f_{i-1}(x) + 1$ for all $x \in \bar{U}_i$.

Since $\bar{V}_i$ and $\bar{U}_i - U_i$ are disjoint closed subsets of a normal space, we can find a map satisfying the stated condition on these two subspaces using the Tietze extension theorem. Taking the minimum of this map and $2f_{i-1}(x) + 1$ produces the desired map λ_i.

Now define ϕ_i by

$$\phi_i(x, t) = \begin{cases} (x, t) & \text{if } \lambda_i(x) \leq t \leq 1 \\ (x, 2t - \lambda_i(x)) & \text{if } f_{i-1}(x) \leq t \leq \lambda_i(x). \end{cases}$$

The behavior of the map ϕ_i may be described as taking each interval $\{x\} \times [\frac{1}{2}(\lambda_i(x) - 1), \lambda_i(x)]$ linearly onto $\{x\} \times [-1, \lambda_i(x)]$, recalling that $\frac{1}{2}(\lambda_i(x) - 1) \leq f_{i-1}(x)$ with equality holding on $\bar{V}_i$ (Figure II.5).

We may now use ϕ_i to alter g_{i-1} to produce g_i. Specifically

$$g_i(x) = \begin{cases} (h_i\phi_i h_i^{-1})g_{i-1}(x) & \text{if } x \in g_{i-1}(M) \cap h_i(\bar{U}_i \times [-1, 1]) \\ g_{i-1}(x) & \text{otherwise.} \end{cases}$$

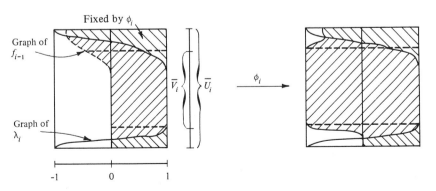

Figure II.5

Both g_i and g_i^{-1} are continuous. We define $f_i \colon B \to [-1, 0]$ by setting $f_i(x) = \pi(g_i(x))$ where π is the projection from $B \times [-1, 0]$ onto the second factor.

This completes the induction step and the homeomorphism g_m gives the required collaring of B. $\square$

REFERENCES

Adams, J. F. [1960]. On the nonexistence of elements of Hopf invariant one. *Ann. of Math.* **72**, 20–104.

Adams, J. F. [1962]. Vector fields on spheres. *Ann. of Math.* **75**, 603–632.

Adem, J. [1952]. The iteration of the Steenrod squares in algebraic topology. *Proc. Nat. Acad. Sci. U.S.A.* **38**, 720–726.

Birkhoff, G. D. [1913]. Proof of Poincaré's geometric theorem. *Trans. Amer. Math. Soc.* **14**, 14–22.

Brown, M. [1962]. Locally flat embeddings of topological manifolds. *In* "Topology of 3-Manifolds and Related Topics" (M. K. Fort, Jr., ed.), pp. 83–91. Prentice-Hall, Englewood Cliffs, New Jersey.

Brown, R. [1968]. "Elements of Modern Topology." McGraw-Hill, New York.

Brown, R. F. [to appear] "The Lefschetz Fixed Point Theorem."

Brown, R. F., and Fadell, E. [1964]. Non-singular path fields on compact topological manifolds. *Notices Amer. Math. Soc.* **11**, 533.

Brumfiel, G., Madsen, I., and Milgram, R. J. [1971]. PL characteristic classes and cobordism. *Bull. Amer. Math. Soc.* **77**, 1025–1030.

Conelly, R. [1971]. A new proof of Brown's collaring theorem. *Proc. Amer. Math. Soc.* **27**, 180–182.

Dold, A. [1965]. Fixed point index and fixed point theorem for Euclidean neighborhood retracts. *Topology* **4**, 1–8.

Eckmann, B. [1943]. Gruppentheoretischer Beweis des Satzes von Hurwitz-Radon über die Komposition quadratischer Formen. *Comment. Math. Helv.* **15**, 358–366.

Eilenberg, S., and Steenrod, N. [1952]. "Foundations of Algebraic Topology." Princeton Univ. Press, Princeton, New Jersey.

Hu, S. T. [1959]. "Homotopy Theory." Academic Press, New York.

Kirby, R., and Siebenmann, L. [1969]. Foundations of TOPology. *Notices Amer. Math. Soc.* **16**, 848.

Lefschetz, S. [1926]. Intersections and transformations of complexes in manifolds. *Trans. Amer. Math. Soc.* **29**, 429–462.

Milnor, J. W. [1957]. Lectures on characteristic classes. Mimeographed notes, Princeton University, Princeton, New Jersey.

Milnor, J. W. [1965]. "Topology from the Differentiable Viewpoint." University of Virginia Press, Charlottesville, Virginia.

Mosher, R. E., and Tangora, M. C. [1968]. "Cohomology Operations and Applications in Homotopy Theory." Harper, New York.

Samelson, H. [1965]. On Poincaré duality. *J. Anal. Math.* **14**, 323–336.

Singer, I. M., and Thorpe, J. A. [1967]. "Lecture Notes on Elementary Topology and Geometry." Scott, Foresman, Glenview, Illinois.

Steenrod, N., and Epstein, D. B. A. [1962]. "Cohomology Operations." Princeton University Press, Princeton, New Jersey.

Stong, R. E. [1968]. "Notes on Cobordism Theory." Princeton University Press, Princeton, New Jersey.

BIBLIOGRAPHY

BOOKS AND NOTES

Abraham, R., and Robbin, J. (1967). "Transversal Mappings and Flows." Benjamin, New York.

Adams, J. F. (1966). "Stable Homotopy Theory" (Lecture Notes in Mathematics), Vol. 3. Springer-Verlag, Berlin and New York.

Aleksandrov, P. S. (1956). "Combinatorial Topology," Vols. 1–4. Graylock Press, Rochester, New York.

Aleksandrov, P. S., and Hopf, H. (1965). "Topologie." Chelsea, New York.

Arkowitz, M., and Curjel, C. R. (1967). "Groups of Homotopy Classes" (Lecture Notes in Mathematics), Vol. 4. Springer-Verlag, Berlin and New York.

Artin, E., and Braun, H. (1969). "Introduction to Algebraic Topology." Merrill Publ., Columbus, Ohio.

Atiyah, M. F. (1967). "K-Theory." Benjamin, New York.

Auslander, L., and MacKenzie, R. E. (1963). "Introduction to Differentiable Manifolds." McGraw-Hill, New York.

Berger, M., and Berger, M. (1968). "Perspectives in Nonlinearity. An Introduction to Nonlinear Analysis." Benjamin, New York.

Bishop, R. L., and Crittenden, R. J. (1964). "Geometry of Manifolds." Academic Press, New York.

Borel, A. (1964). "Cohomologie des éspaces localement compacts d'après J. Leray" (Lecture Notes in Mathematics), Vol. 2. Springer-Verlag, Berlin and New York.

Borel, A. (1967). "Topics in the Homology Theory of Fibre Bundles" (Lecture Notes in Mathematics), Vol. 36. Springer-Verlag, Berlin.

Bott, R. (1969). "Lectures on $K(X)$." Benjamin, New York.

Bourgin, D. G. (1963). "Modern Algebraic Topology." Macmillan, New York.

Bredon, G. E. (1967). "Equivariant Cohomology Theories" (Lecture Notes in Mathematics), Vol. 34. Springer-Verlag, Berlin and New York.

Bredon, G. E. (1967). "Sheaf Theory." McGraw-Hill, New York.

Bredon, G. E. (1972). "An Introduction to Compact Transformation Groups." Academic Press, New York.

Brinkmann, H. B., and Puppe, D. (1966). "Kategorien und Funktoren" (Lecture Notes in Mathematics), Vol. 18. Springer-Verlag, Berlin and New York.

Bröcker, T., and tom Dieck, T. (1970). "Kobordismentheorie" (Lecture Notes in Mathematics), Vol. 178. Springer-Verlag, Berlin and New York.

Browder, W. (1972). "Surgery on Simply Connected Manifolds." Springer-Verlag, Berlin and New York.

Brown, R. (1968). "Elements of Modern Topology." McGraw-Hill, New York.

Cairns, S. S. (1961). "Introductory Topology." Ronald Press, New York.

Cartan, H. (1948). "Algebraic Topology" (G. Springer and H. Pollak, eds.). Lecture notes, Harvard Univ., Cambridge, Massachusetts.

Cartan, H., and Eilenberg, S. (1956). "Homological Algebra." Princeton Univ. Press, Princeton, New Jersey.

"Category Theory, Homology Theory and Their Applications I, II and III" (1969). (Lecture Notes in Mathematics), Vols. 86, 92, 99. Springer-Verlag, Berlin and New York.

Cerf, J. (1968). "Sur les difféomorphismes de la sphere de dimension trois ($\Gamma_4 = 0$)" (Lecture Notes in Mathematics), Vol. 53. Springer-Verlag, Berlin and New York.

Chevalley, C. (1960). "Lie Groups." Princeton, Univ. Press. Princeton, New Jersey.

"Colloque de Topologie." (1966). Tenu à Bruxelles du 7 au 10 Septembre 1964. Gauthier-Villars, Paris.

"Colloque de Topologie Différentielle" (1969). Univ. de Montpellier, Montpellier.

"Colloquium on Algebraic Topology" (1962). Aarhus Univ., Aarhus.

"Conference on Algebraic Topology" (1968). Univ. of Illinois at Chicago Circle, Chicago, Illinois.

Conner, P. E. (1967). "Seminar on Periodic Maps" (Lecture Notes in Mathematics), Vol. 46. Springer-Verlag, Berlin and New York.

Conner, P. E. (1968). "Lectures on the Action of a Finite Group" (Lecture Notes in Mathematics), Vol. 73. Springer-Verlag, Berlin.

Conner, P. E., and Floyd, E. E. (1964). "Differentiable Periodic Maps," Ergebnisse der Mathematik und ihrer Grenzgebiete, N. F. Vol. 33. Springer-Verlag, Berlin and New York.

Conner, P. E., and Floyd, E. E. (1966). "The Relation of Cobordism to K-Theories" (Lecture Notes in Mathematics), Vol. 28. Springer-Verlag, Berlin and New York.

Cooke, G. E., and Finney, R. L. (1967). "Homology of Cell Complexes" (based on lectures of N. Steenrod). Princeton Univ. Press, Princeton, New Jersey.

Crowell, R. H., and Fox, R. H. (1963). "Introduction to Knot Theory." Ginn, Boston, Massachusetts.

Curtis, E. (1967). "Simplicial Homotopy Theory." Math. Inst., Aarhus Univ., Aarhus.

de Rham, G. (1960). "Variétés Différentiables." Hermann, Paris.

de Rham, G., Maumary, S., and Kervaire, M. A. (1967). "Torsion et Type Simple d'Homotopie" (Lecture Notes in Mathematics), Vol. 48. Springer-Verlag, Berlin and New York.

"Differential and Combinatorial Topology. A Symposium in Honor of Marston Morse" (1965). Princeton Univ. Press, Princeton, New Jersey.

Dold, A. (1966). "Halbexakte Homotopiefunktoren" (Lecture Notes in Mathematics), Vol. 12. Springer-Verlag, Berlin and New York.

Dold, A. (1968). "On General Cohomology," Chapters 1–9. Math. Inst., Aarhus Univ., Aarhus.

Dugundji, J. (1960). "Topology." Allyn & Bacon, Boston, Massachusetts.

Dyer, E. (1969). "Cohomology Theories." Benjamin, New York.

Eckmann, B. (1962). "Homotopie et cohomologie." Presses de l'Univ. de Montréal, Montréal, Quebec.

Eells, J., Jr. (1967). "Singularities of Smooth Maps." Gordon & Breach, New York.

Eilenberg, S., and Steenrod, N. (1952). "Foundations of Algebraic Topology." Princeton Univ. Press, Princeton, New Jersey.

"Essays on Topology and Related Topics. Mémoires dédiés à Georges de Rham" (1970). Springer-Verlag, Berlin and New York.

Franz, W. (1965). Topologie II. Algebraische Topologie." Walter de Gruyter, Berlin.

Fréchet, M., and Fan, K. (1967). "Introduction to Combinatorial Topology." Prindle, Weber, & Schmidt, Boston, Massachusetts.

Fuks, D. B., and Fomenko, A. (1967–1968). "Homotopic Topology," Vols. I and II (in Russian). Izdat. Moskow Univ., Moscow.

Gabriel, P., and Zisman, M. (1967). "Calculus of Fractions and Homotopy Theory" (Ergebnisse der Mathematik und ihre Grenzgebiete), Vol. 35. Springer-Verlag, Berlin and New York.

"Global Analysis" (1970). *Proc. Symp. Pure Math.* Vols. 14–16. Amer. Math. Soc., Providence, Rhode Island.

"Global Analysis. Papers in Honor of K. Kodaira" (1969). Univ. of Tokyo Press, Tokyo.

Godement, R. (1958). "Topologie Algébrique et Théorie des Faisceaux" (Publ. Math. Univ. Strasbourg), No. 13. Hermann, Paris.

Goldberg, S. (1962). "Curvature and Homology." Academic Press, New York.

Greenberg, M. (1967). "Lectures on Algebraic Topology." Benjamin, New York.

Grothendieck, A. (1955). "A General Theory of Fibre Spaces with Structure Sheaf." Univ. of Kansas, Lawrence, Kansas.

Hilbert, D., and Cohn-Vossen, S. (1932). "Anschauliche Geometrie." Springer-Verlag, Berlin and New York. [English translation: Hilbert, D., and Cohn-Vossen, S. (1952). "Geometry and the Imagination." Chelsea, New York.]

Hilton, P. (1953). "An Introduction to Homotopy Theory" (Cambridge Tracts in Mathematics), No. 43. Cambridge Univ. Press, London and New York.

Hilton, P. J. (1965). "Homotopy Theory and Duality." Gordon & Breach, New York.

Hilton, P. J. (ed.) (1968). "Studies in Modern Topology" (Studies in Mathematics), Vol. 5. Prentice-Hall, Englewood Cliffs, New Jersey.

Hilton, P. J. (ed.) (1971). "Symposium on Algebraic Topology, Battelle—1971" (Lecture Notes in Mathematics), Vol. 249. Springer-Verlag, Berlin and New York.

Hilton, P. J., and Wylie, S. (1960). "Homology Theory." Cambridge Univ. Press, London and New York.

Hirzebruch, F. (1965). "New Topological Methods in Algebraic Geometry." Springer-Verlag, Berlin and New York.

Hirzebruch F., and Mayer, K. H. (1968). "$O(n)$-Mannigfaltigkeiten, exotische Sphären und Singularitäten" (Lecture Notes in Mathematics), Vol. 57. Springer-Verlag, Berlin and New York.

Hocking, J. G., and Young, G. S. "Topology." Addison-Wesley, Reading, Massachusetts.

Hu, S.-T. (1959). "Homotopy theory" (Pure and Appl. Mathematics), Vol. 8. Academic Press, New York.

Hu, S.-T. (1965). "Theory of Retracts." Wayne State Univ. Press, Detroit, Michigan.

Hu, S.-T. (1966). "Homology Theory: A First Course in Algebraic Topology." Holden-Day, San Francisco, California.

Hu, S.-T. (1968). "Cohomology Theory." Markham Publ., Chicago, Illinois.

Hu. S.-T. (1969). "Differentiable Manifolds." Holt, New York.

Hudson, J. F. P. (1969). "Piecewise Linear Topology." Benjamin, New York.

Hurewicz, W., and Wallman, H. (1941). "Dimension Theory." Princeton Univ. Press, Princeton, New Jersey.

Husemoller, D. (1966). "Fibre Bundles." McGraw-Hill, New York.

Husseini, S. Y. (1969). "The Topology of Classical Groups and Related Topics." Gordon & Breach, New York.

"International Symposium on Algebraic Topology" (1958). Univ. Nacional Automata de México, México, D.F.

Janich, K. (1968). "Differenzierbare G-Mannigfaltigkeiten" (Lecture Notes in Mathematics), Vol. 59. Springer-Verlag, Berlin and New York.

Kamber, F., and Tondeur, P. (1968). "Flat Manifolds," (Lecture Notes in Mathematics), Vol. 67. Springer-Verlag, Berlin and New York.

Kerékjártó, B. (1923). "Vorlesungen über Topologie" (Grundlehren der Math. Wissenschaften), Vol. 8. Springer-Verlag, Berlin and New York.

Kultze, R. (1970). "Garbentheorie." Teubner, Stuttgart.

Lamotke, K. (1968). "Semisimpliziale Algebraische Topologie" (Die Grundlehren der Mathematischen Wissenschaften), Vol. 147. Springer-Verlag, Berlin and New York.

Lang, S. (1962). "Introduction to Differentiable Manifolds." Wiley (Interscience), New York.

"Lectures given at Nordic Summer School in Mathematics" (1968). Math. Inst., Aarhus Univ., Aarhus.

Lefschetz, S. (1930). "Topology" (Colloq. Publ.), Vol. 12. Amer. Math. Soc. New York.

Lefschetz, S. (1942). "Algebraic Topology" (Colloq. Publ.), Vol. 27. Amer. Math. Soc., New York.

Lefschetz, S. (1942). "Topics in Topology," Annals of Math. Stud., No. 10. Princeton Univ. Press, Princeton, New Jersey.

Lefschetz, S., (1949). "Introduction to Topology." Princeton Univ. Press, Princeton, New Jersey.

Maclane, S. "Homology." Academic Press, New York and Springer-Verlag, Berlin and New York.

"Manifolds—Amsterdam 1970" (1971). (Lecture Notes in Mathematics), Vol. 197. Springer-Verlag, Berlin and New York.

Massey, W. S. (1967). "Algebraic Topology: An Introduction." Harcourt, New York.

May, J. P. (1967). "Simplicial Objects in Algebraic Topology." Van Nostrand–Reinhold, Princeton, New Jersey.

Mayer, K. H. (1969). "Relationen zwischen characteristischen Zahlen" (Lecture Notes in Mathematics), Vol. 111. Springer-Verlag, Berlin and New York.

Milnor, J. (1957). Lectures on characteristic classes. Mimeographed notes, Princeton University, Princeton, New Jersey.

Milnor, J. (1958). "Lectures on Differential Topology," mimeographed notes. Princeton Univ., Princeton, New Jersey.

Milnor, J. (1963). "Morse Theory" (Annals of Mathematics Studies), No. 51. Princeton Univ. Press, Princeton, New Jersey.

Milnor, J. (1965). "Lectures on the *h*-Cobordism Theorem" (notes by L. Siebenmann and J. Sondow). Princeton Univ., Princeton, New Jersey.

Milnor, J. (1968). "Singular Points of Complex Hypersurfaces" (Annals of Mathematics Stud.), No. 61. Princeton Univ., Princeton, New Jersey.

Morse, M., and Cairns, S. S. (1969). "Critical Point Theory in Global Analysis and Differential Geometry: An Introduction." Academic Press, New York.

Munkres, J. R. (1966). "Elementary Differential Topology" (Annals of Mathematics Stud.), No. 54. Princeton Univ. Press, Princeton, New Jersey.

Nagano, T. (1970). "Homotopy Invariants in Differential Geometry" (Memoires of Amer. Math. Soc.), No. 100, Providence, Rhode Island.

Narasimhan, M. S., Ramanan, S., Sridharan, R., and Varadarajan, K. (1964). "Algebraic Topology." Tata Inst. of Fundamental Res., Bombay.

Nomizu, K. (1956). "Lie Groups and Differential Geometry." Math. Soc. of Japan, Tokyo.

Northcott, D. G. (1960). "An Introduction to Homological Algebra." Cambridge Univ. Press, London and New York.

Palais, R. S. (1968). "Foundations of Global Nonlinear Analysis." Benjamin, New York.

Patterson, E. M. (1956). "Topology." Wiley (Interscience), New York.

Poincaré, H. (1895). Analysis Situs. *J. École Poly. Paris* (2) **1**, 1–123.

Pontrjagin, L. S. (1959) Smooth manifolds and their applications in homotopy theory, *Amer. Math. Soc. Transl., Ser. 2*, **11**, 1–114.

Porteous, I. R. (1969). "Topological Geometry." Van Nostrand-Reinhold, Princeton, New York.

Proc. Liverpool Singularities Symp. I, 1969/70 (1971). (Lecture Notes in Mathematics), Vol. 192. Springer-Verlag, Berlin and New York.

Proc. Advan. Study Inst. Algebraic Topol. (1970). Mat. Inst., Aarhus Univ., Aarhus.

Proc. Conf. Algebraic Topol. Madison, 1970 (1971). *Proc. Symp. Pure Math.* **22.** Amer. Math. Soc., Providence, Rhode Island.

Proc. Conf. Transformation Groups, New Orleans, 1967 (1968). Springer-Verlag, Berlin and New York.

Quillen, D. G. (1967). "Homotopical Algebra" (Lecture Notes in Mathematics), Vol. 43. Springer-Verlag, Berlin and New York.

"Reviews of Papers in Algebraic and Differential Topology, Topological Groups and Homological Algebra" (1968). Vols. I and II (compiled by N. Steenrod). Amer. Math. Soc., Providence, Rhode Island.

Seifert, H., and Threlfall, W. (1934). "Lehrbuch der Topologie." Teubner, Berlin.

Séminaire Heidelberg-Strasbourg, 1966/67 (1969). "Dualité de Poincaré." Univ. de Strasbourg, Strasbourg.

Séminaire Henri Cartan de l'École Normale Superieure (publ. annually beginning in 1948–49). Secrétariat Mathématique, Paris.

"Seminar on Combinatorial Topology" (1963). Inst. Hautes Études Sci., Paris.

Sigrist, F. (ed.) (1971). "*H*-Spaces-Neuchâtel (Suisse)" (Lecture Notes in Mathematics), Vol. 196. Springer-Verlag, Berlin and New York.

Singer, I. M., and Thorpe, J. A. (1961). "Lecture Notes on Elementary Topology and Geometry." Scott, Foresman, Glenview, Illinois.

Spanier, E. H. (1966). "Algebraic Topology." McGraw-Hill, New York.

Spivak, M. (1965). "Calculus on Manifolds." Benjamin, New York.

Spivak, M. (1971). "Differential Geometry," Vols. I and II. Publish or Perish, Cambridge, Massachusetts.

Stallings, J. R. (1967). "Lectures on Polyhedral Topology." Tata Inst. of Fundamental Res., Bombay.

Stasheff, J. (1970). "*H*-Spaces from the Homotopy Point of View" (Lecture Notes in Mathematics), Vol. 161. Springer-Verlag, Berlin and New York.

"The Steenrod Algebra and its Applications: A Conference to Celebrate N. Steenrod's Sixtieth Birthday" (1970). (F. Peterson, ed.) (Lecture Notes in Mathematics), Vol. 168. Springer-Verlag, Berlin and New York.

Steenrod, N. (1951). "The Topology of Fibre Bundles." Princeton Univ. Press, Princeton, New Jersey.

Steenrod, N. E., and Epstein, D. B. A. (1962). "Cohomology Operations" (Annals of Mathematics Studies), No. 50. Princeton Univ. Press, Princeton, New Jersey.

Stong, R. E. (1968). "Notes on Cobordism Theory." Princeton Univ. Press, Princeton, New Jersey.

Teleman, C. (1968). "Grundzüge der Topologie und Differenzierbare Mannigfaltigkeiten," (Mathematische Monographien), Vol. 8. VEB Deutscher Verlag der Wissenschaften, Berlin.

Thomas, E. (1966). "Seminar on Fiber Spaces" (Lecture Notes in Mathematics), Vol. 13. Springer-Verlag, Berlin and New York.

Toda, H. (1962). "Composition Methods in Homotopy Groups of Spheres" (Annals of Mathematics Studies), No. 49. Princeton Univ. Press, Princeton, New Jersey.

tom Dieck, T., Kamps, K. H., and Puppe, D. (1970). "Homotopietheorie" (Lecture Notes in Mathematics), Vol. 157. Springer-Verlag, Berlin and New York.

Tondeur, P. (1965). "Introduction to Lie Groups and Transformation Groups" (Lecture Notes in Mathematics), Vol. 7. Springer-Verlag, Berlin and New York.

"Topology of Manifolds, Georgia—1969" (1970). Markham Publ., Chicago, Illinois.

"Topology of 3-Manifolds and Related Topics, Georgia—1961" (1962). Prentice-Hall, Englewood Cliffis, New Jersey.

"Topology Seminar, Wisconsin—1965" (1966). Ann. of Math. Studies, No. 60. Princeton Univ. Press, Princeton, New Jersey.

Veblen, O. (1931). "Analysis Situs," Vol. 5. Amer. Math. Soc. Colloq. Publ., New York.

Wall, C. T. C. (1970). "Surgery on Compact Manifolds." Academic Press, London and New York.

Wallace, A. H. (1970). "Algebraic Topology: Homology and Cohomology." Benjamin, New York.

Wallace, A. H. (1957). "An Introduction to Algebraic Topology." Pergamon, Oxford.

Wallace, A. H. (1968). "Differential Topology: First Steps." Benjamin, New York.

Whitehead, G. W. (1966). "Homotopy Theory." MIT Press, Cambridge, Massachusetts.

Whitney, H. (1957). "Geometric Integration Theory." Princeton Univ. Press, Princeton, New Jersey.

Wilder, R. L. (ed.) (1941). "Lectures in Topology." Univ. of Michigan Press, Ann Arbor, Michigan.

Wilder, R. L. (1963). "Topology of Manifolds." Vol. 32. Amer. Math. Soc., Providence, Rhode Island.

Wu, Wen-Tsün (1965). "A Theory of Imbedding, Immersion and Isotopy of Polytopes in Euclidean Space." Science Press, Peking.

SURVEY AND EXPOSITORY ARTICLES

Adams, J. F. (1968). A survey of homotopy theory. *Proc. Internat. Congr. Math., Moscow, 1966* 33–43. Izdat. "Mir," Moscow.

Adams, J. F. (1969). "Lectures on Generalized Cohomology" (Lecture Notes Mathematics), Vol. 99, pp. 1–138. Springer-Verlag, Berlin and New York.

Adams, J. F. (1971). Algebraic topology in the last decade. *Proc. Symp. Pure Math.* **22** Amer. Math. Soc., Providence, Rhode Island.

Atiyah, M. F. (1963). The Grothendieck ring in geometry and topology. *Proc. Internat. Congr. Math., Stockholm, 1962* 442–446. Inst. Mittag-Leffler, Djursholm.

Atiyah, M. F. (1966). The role of algebraic topology in mathematics. *J. London Math. Soc.* **41**, 63–69.

Atiyah, M. F. (1967). Algebraic topology and elliptic operators. *Comm. Pure Appl. Math.* **20**, 237–249.

Atiyah, M. F. (1968). Global aspects of elliptic differential operators. *Proc. Internat. Congr. Math., Moscow, 1966* 57–64. Izdat. "Mir," Moscow.

Atiyah, M. F. (1969). Algebraic topology and operators in Hilbert space. *Lectures Modern Anal. Appl.* **1**, 101–121. Springer-Verlag, Berlin and New York.

Atiyah, M. F. (1970). Vector fields on manifolds. "Arbeitsgemeinschaft für Forschung des Landes Nordrhein-Westfalen, Heft 200." Westdeutscher Verlag, Cologne.

Atiyah, M. F., and Hirzebruch, F. (1961). Charakteristische Klassen und Anwendungen. *Enseignement Math.* (2) **7**; 188–213 (1962).

Baum, P. F. (1970). Vector fields and Gauss-Bonnet. *Bull. Amer. Math. Soc.* **76**, 1202–1211.

Bing, R. H. (1967). Challenging conjectures. *Amer. Math. Monthly* **74**, 56–64.

Bognár, M. (1967). The origin and development of the theory of topological manifolds (in Hungarian) *Mat. Lapok* **18**, 37–57.

Borsuk, K. (1958). Über einige Probleme der anschaulichen Topologie. *Jber. Deutsch. Math. Verein.* **60**, 101–114.

Bott, R. (1961). Vector fields on spheres and allied problems. *Enseignement Math.* (2) **7**; 125–138 (1962).

Bott, R. (1970). The periodicity theorem for the classical groups and some of its applications. *Advan. Math.* **4**, 353–411.

Browder, W. (1968). Surgery and the theory of differentiable transformation groups. *Proc. Conf. Transformation Groups* 1–46. Springer-Verlag, Berlin and New York.

Brown, E. H. (1969). The Arf invariant of a manifold. *Conf. Algebraic Topol., 1968* 9–18. Univ. of Illinois at Chicago Circle, Chicago, Illinois.

Brown, R. F. (1971). Notes on Leray's index theory. *Advan. Math.* **7**, 1–28.

Cairns, S. S. (1946). The triangulation problem and its role in analysis. *Bull. Amer. Math. Soc.* **52**, 545–571.

Calugareanu, G. (1970). Points de vue sur la théorie des noeuds. *Enseignement Math.* (2) **16**, 97–110.

Cartan, H. (1968). L'oeuvre de Michael F. Atiyah. *Proc. Internat. Congr. Math., Moscow 1966* 9–14. Izdat. "Mir," Moscow.

Cartan, H. (1970). Les travaux de G. de Rham sur les variétés différentiables. "Essays on Topology and Related Topics," pp. 1–11. Springer-Verlag, Berlin and New York.

Cartan, H. (1970). Structural stability of differential mappings. *Proc. Internat. Conf. Functional Anal. Related Topics, Tokyo, 1969* 1–10. Univ. of Tokyo Press, Tokyo.

Černavskii, A. V. (1963). Geometric topology of manifolds (in Russian). "Itogi Nauki" (Algebraic Topology, 1962), pp. 161–187. Akad. Nauk SSSR, Moscow.

Chern, S. S. (1946). Some new viewpoints in differential geometry in the large. *Bull. Amer. Math. Soc.* **52**, 1–30.

Chern, S. S. La géométrie des sous-variétés d'un éspace euclidien a plusieurs dimensions. *Enseignement Math.* **40**, 26–46.

Chern, S. S. (1966). The geometry of *G*-structures. *Bull. Amer. Math. Soc.* **72**, 167–219.

Chevalley, C., and Weil, A. (1957). Hermann Weyl (1885–1955). *Enseignement Math.* **3**, 157–187.

Čisang, H., and Černavskii, A. V. (1967). Geometric topology of manifolds (in Russian). "Algebra, Topology, Geometry, 1965," pp. 219–261. Akad. Nauk SSSR, Moscow.

Cockcroft, W. H., and Jarvis, T. (1964). An introduction to homotopy theory and duality. I. *Bull. Soc. Math. Belg.* **16**, 407–428; II, **17** (1965), 3–26.

Connell, E. H. (1970). Characteristic classes. *Illinois J. Math.* **14**, 496–521.

de Rham, G. (1967). Introduction aux polynômes d'un noeud. *Enseignement Math.* (2) **13**; 187–194 (1968).

de Rham, G. (1961). La théorie des formes différentielles extérieures et l'homologie des variétés différentiables. *Rend. Mat. e Appl.* (5) **20**, 105–146.

de Rham, G. (1965). Sur les invariants de torsion de Reidemeister-Franz, et de J. H. C. Whitehead. *Rev. Roumaine Math. Pure Appl.* **10**, 679–685.

Dold, A. (1968). (Co-)homology properties of topological manifolds. *Conf. Topol. Manifolds* 47–57. Prindle, Weber, and Schmidt, Boston, Massachusetts.

Dold, A. (1971). Chern classes in general cohomology. *in* "Symposia Mathematica," Vol. 5. Academic Press, London and New York.

Duke, R. A. (1970). Geometric embedding of complexes. *Amer. Math. Monthly* **77**, 597–603.

Eckmann, B. (1962). Homotopie und Homologie. *Enseignement Math.* (2) **8**, 209–217.

Eckmann, B. (1971). Simple homotopy type and categories of fractions. *Symp. Math.* **5**, 285–299.

Eilenberg, S. (1949). On the problems of topology. *Ann. of Math.* **50**, 247–260.

Eilenberg, S. (1963). Algebraic topology. "Lectures on Modern Mathematics," Vol. I, pp. 98–114. Wiley, New York.

Eisenack, G. (1969). Lefschetzche Fixpunkträume und Fixpunkte von iterierten Abbildungen." Gesellschaft für Math. und Datenverarbeitung, Bonn.

Fadell, E. (1970). Recent results in the fixed point theory of continuous maps. *Bull. Amer. Math. Soc.* **76**, 10–29.

Fuks, D. B. (1969). Homotopic topology (in Russian), "Algebra, Topology, Geometry, 1969," pp. 71–122. Akad. Nauk SSSR, Moscow.

Glezerman, M., and Pontrjagin, L. (1951). Intersections in manifolds. *Amer. Math. Soc. Transl.* **50**, 149 pp.

Haefliger, A., and Reeb, G. (1957). Variétés à une dimension et structures feuilletées du plan. *Enseignement Math.* (2) **3**, 107–125.

Haken, W. (1970). Various aspects of the three-dimensional Poincaré problem. "Topology of Manifolds," pp. 140–152. Markham, Chicago, Illinois.

Hilton, P. J. (1961). Memorial tribute to J. H. C. Whitehead. *Enseignement Math.* (2) **7**, 107–124.

Hirsch, G. (1964). Introduction à la topologie algébrique. I. *Bull. Soc. Math. Belg.* **16**, 152–188.

Hirsch, G. (1966). Introduction à la topologie algébrique. II. Les groups d'homotopie. *Bull. Soc. Math. Belg.* **18**, 227–244.

Hirzebruch, F. (1954). Some problems on differentiable and complex manifolds. *Ann. of Math.* **60**, 213–236.

Hirzebruch, F. (1966). Über singularitäten komplexer Flächen. *Rend. Mat. e Appl.* **25**, 213–232.

Holm, P. (1967). Microbundles and *S*-duality. *Acta Math.* **118**, 271–296.

Hopf, H. (1950). Die *n*-dimensionalen Sphären und projektiven Räume in der Topologie. *Proc. Internat. Congr. Math. Cambridge* **1**, 193–202.

Hopf, H. (1951). Über komplex-analytische Mannigfaltigkeiten. *Univ. Roma. Inst. Naz. Alta Mat. Rend. Mat. e Appl.* **10**, 169–182.

Hopf, H. (1952). Einige Anwendungen der Topologie auf die Algebra. Univ. e Politecnico Torino. *Rend. Sem. Mat.* **11**, 75–91.

Hopf, H. (1953). Vom Bolzanoschen Nullstellensatz zur algebraischen Homotopie-theorie der Sphären. *Jber. Deutsch. Math. Verein.* **56**, 59–76.

Hopf, H. (1962). Über den Defekt stetiger Abbildungen von Mannigfaltigkeiten. *Rend. Mat. e Appl.* (5) **21**, 273–285.

Hopf, H. (1966). Ein Abschnitt aus der Entwicklung der Topologie. *Jber. Deutsch. Math. Verein.* **68**, 182–192.

Hu, S.-T. (1947). An exposition of the relative homotopy theory. *Duke Math. J.* **14**, 991–1033.

Hurewicz, W. (1950). Homotopy and homology. *Proc. Internat. Cong. Math. Cambridge* **2**, 344–349.

James, I. M. (1971). "Bibliography on *H*-Spaces" (Lecture Notes in Mathematics), Vol. 196, pp. 137–156. Springer-Verlag, Berlin and New York.

Lashof, R. (1965). Problems in differential and algebraic topology. *Ann. of Math.* **81**, 565–591.

Lee, C. N. (1968). Equivariant homology theories. *Proc. Conf. Transformation Groups* 237–244. Springer-Verlag, Berlin and New York.

Lefschetz, S. (1968). A page of mathematical autobiography. *Bull. Amer. Math. Soc.* **74**, 854–879.

Mahowald, M. E. (1968). Some remarks on obstruction theory. *Conf. Topological Manifolds* 107–113. Prindle, Weber, and Schmidt, Boston, Massachusetts.

Massey, W. S. (1954). Some new algebraic methods in topology. *Bull. Amer. Math. Soc.* **60**, 111–123.

Massey, W. S. (1955). Some problems in algebraic topology and the theory of fibre bundles. *Ann. of Math.* **62**, 327–359.

Milnor, J. (1961). A procedure for killing homotopy groups of differentiable manifolds, *Symp. Pure Math.* **3**, 39–55, Differential Geometry. Amer. Math. Soc., Providence, Rhode Island.

Milnor, J. (1961). Differentiable manifolds with boundary (in Spanish). *Ann. Inst. Mat. Univ. Nac. Autónoma México* **1**, 82–116.

Milnor, J. (1962). A survey of cobordism theory. *Enseignement Math.* (2) **8**, 16–23.

Milnor, J. (1963). Spin structures on manifolds. *Enseignement Math.* (2) **9**, 198–203.

Milnor, J. (1964). "Differential Topology" (Lectures on Modern Mathematics), Vol. 2, pp. 165–183. Wiley, New York.

Milnor, J. (1966). Whitehead torsion. *Bull. Amer. Math. Soc.* **72**, 358–426.

Milnor, J. (1968). Infinite cyclic coverings. *Conf. Topological Manifolds* 115–133. Prindle, Weber & Schmidt, Boston, Massachusetts.

Milnor, J., and Burlet, O. (1970). Torsion et type simple d'homotopie. "Essays on Topology and Related Topics," pp. 12–17. Springer Publ., New York.

Misner, C. (1964). Differential geometry and differential topology. "Relativité, Groupes et Topologie" (Univ. Grenoble-1963), pp. 881–929. Gordon & Breach, New York.

Munkres, J. R. (1967). Concordance of differentiable structures—Two approaches. *Mich. Math. J.* **14**, 183–191.

Newman, M. H. A. (1963). Geometrical topology. *Proc. Internat. Congr. Math., Stockholm, 1962* 139–146. Inst. Mittag-Leffler, Djursholm.

Novikov, S. P. (1963). Differential topology (in Russian). "Itogi Nauki" (Algebraic Topology, 1962). pp. 134–160. Akad. Nauk SSSR, Moscow.

Novikov, S. P. (1965). New ideas in algebraic topology. *K*-Theory and its applications. *Russian Math. Surveys* **20**, No. 3, 37–62.

Novikov, S. P. (1966). The Cartan-Serre theorem and inner homologies (in Russian). *Uspehi Mat. Nauk* **21**, No. 5 (131), 217–232.

Novikov, S. P. (1968). Pontrjagin classes, the fundamental group and some problems of stable algebra. *Amer. Math. Soc. Transl.* **70**, 172–179.

Novikov, S. P. (1968). Rational Pontrjagin classes. Homeomorphism and homotopy type of closed manifolds. I. *Amer. Math. Soc. Transl.* **66**, 214–230.

Olum, P. (1969). Factorizations and induced homomorphisms. *Advan. Math.* **3**, 72–100.

Palais, R. S. (1970). Critical point theory and the minimax principle. "Global Analysis" (*Proc. Symp. Pure Math.*), Vol. 15, pp. 185–212. Amer. Math. Soc., Providence, Rhode Island.

Papakyriakopoulos, C. D. (1958). Some problems on 3-dimensional manifolds. *Bull. Amer. Math. Soc.* **64**, 317–335.

Papakyriakopoulos, C. D. (1958). The theory of 3-dimensional manifolds since 1950. *Proc. Internat. Congr. Math.* 433–440.

Peterson, F. P. (1968). "Lectures on Cobordism Theory," Lectures in Mathematics, Kyoto Univ., Vol. 1. Kinokuniya Book Store, Tokyo.

Peterson, F. P. (1969). Characteristic classes—Old and new. *Bull. Amer. Math. Soc.* **75**, 907–911.

Quillen, D. (1971). Elementary proofs of some results of cobordism theory using Steenrod operations. *Advan. Math.* **7**, 29–56.

Reid, W. T. (1969). Remarks on the Morse index theorem. *Proc. Amer. Math. Soc.* **20**, 339–341.

Rohlin, V. A. (1963). Theory of intrinsic homologies. *Amer. Math. Soc. Transl.* (2) **30**, 255–271.

Roussarie, R. (1969). Sur les fuilletages des variétés orientables de dimension 3. *Colloque de Topolog. Différentielle, 1969* 50–63. Univ. de Montpellier, Montpellier.

Seebach, J. A., Seebach, L. A., and Steen, L. A. (1970). What is a sheaf? *Amer. Math. Monthly* **77**, 681–703.

Segal, G. (1968). Equivariant *K*-Theory. *Inst. Hautes Études Sci. Publ. Math.* No. 34, 129–151.

Shih, W. (1968). Characteristic classes as natural transformations and topological index of classical elliptic operators. *Cahiers Topolog. Géom. Différentielle* **10**, 395–447.

Smale, S. (1963). A survey of some recent developments in differential topology. *Bull. Amer. Math. Soc.* **69**, 131–145.

Smale, S. (1967). Differentiable dynamical systems. *Bull. Amer. Math. Soc.* **73**, 747–817.

Smale, S. (1968). Differentiable dynamical systems. *Proc. Internat. Congr. Math. Moscow, 1966* 139. Izdat. "Mir," Moscow.

Smale, S. (1969). Global stability questions in dynamical systems. "Lectures in Modern Analysis and Applications," Vol. I, pp. 150–158. Springer-Verlag, Berlin and New York.

Smale, S. (1969). What is global analysis? *Amer. Math. Monthly* **76**, 4–9.

Smale, S. (1970). Notes on differentiable dynamical systems. "Global Analysis" (*Proc. Symp. Pure Math.*), Vol. 14, pp. 277–287. Amer. Math. Soc., Providence, Rhode Island.

Stasheff, J. (1971). "Infinite Loop Spaces—A Historical Survey" (Lecture Notes in Mathematics), Vol. 196, pp. 43–53. Springer-Verlag, Berlin and New York.

Steenrod, N. E. (1957). The work and influence of Prof. S. Lefschetz in algebraic topology. "Algebraic Geometry and Topology," pp. 24–43. Princeton Univ. Press, Princeton, New Jersey.

Steenrod, N. E. (1961) The cohomology algebra of a space. *Enseignement Math.* **7**, 153–178.

Steenrod, N. E. (1972). Cohomology operations and obstructions to extending continuous functions. *Advan. Math.* **8**, 371–416.

Sugawara, M. (1968). Introduction to *H*-spaces (in Japanese). *Sûgaku* **20**, 202–211.

Thom, R. (1968). Sur les travaux de Stephen Smale. *Proc. Internat. Congr. Math., Moscow, 1966* 25–28. Izdat. "Mir," Moscow.

Thom, R. (1969). Ensembles et morphismes stratifiés. *Bull. Amer. Math. Soc.* **75**, 240–284.

Thomas, E. (1969). Vector fields on manifolds. *Bull. Amer. Math. Soc.* **75**, 643–683.

Tucker, A. W. (1945). Some topological properties of disk and sphere. *Proc. Canad. Math. Congr., 1st, Montreal, 1945* 285–309.

Wall, C. T. C. (1965). Topology of smooth manifolds. *J. London Math. Soc.* **40**, 1–20.

Wall, C. T. C. (1968). Homeomorphism and diffeomorphism classification of manifolds. *Proc. Internat. Congr. Math. Moscow, 1966* 450–460. Izdat. "Mir," Moscow.

Wallace, A. H. (1962). A geometric method in differential topology. *Bull. Amer. Math. Soc.* **68**, 533–542.

Weber, C. (1967). Quelques théorèmes bien connus sur les ANR et les CW complexes. *Enseignement Math.* (2) **13**, 211–220.

Whitehead, G. W. (1950). Homotopy groups of spheres. *Proc. Internat. Congr. Math. Cambridge, 1950* **2**, 358–362.

Whitehead, G. W. (1970). "The Work of Norman E. Steenrod in Algebraic Topology: An Appreciation" (Lecture Notes in Mathematics), Vol. 168, pp. 1–10. Springer-Verlag, Berlin and New York.

Whitehead, J. H. C. (1956). Duality in topology. *J. London Math. Soc.* **31**, 134–148.

Whittlesey, E. F. (1963). Fixed points and antipodal points. *Amer. Math. Monthly* **70**, 807–821.

Wu, Wen-Tsün (1959). Topologie combinatoire et invariants combinatoire. *Colloq. Math.* **7**, 1–8.

INDEX

Pure and Applied Mathematics

A Series of Monographs and Textbooks

Editors **Samuel Eilenberg and Hyman Bass**

Columbia University, New York

RECENT TITLES

XIA DAO-XING. Measure and Integration Theory of Infinite-Dimensional Spaces: Abstract Harmonic Analysis

RONALD G. DOUGLAS. Banach Algebra Techniques in Operator Theory

WILLARD MILLER, JR. Symmetry Groups and Their Applications

ARTHUR A. SAGLE AND RALPH E. WALDE. Introduction to Lie Groups and Lie Algebras

T. BENNY RUSHING. Topological Embeddings

JAMES W. VICK. Homology Theory: An Introduction to Algebraic Topology

E. R. KOLCHIN. Differential Algebra and Algebraic Groups

GERALD J. JANUSZ. Algebraic Number Fields

A. S. B. HOLLAND. Introduction to the Theory of Entire Functions

WAYNE ROBERTS AND DALE VARBERG. Convex Functions

A. M. OSTROWSKI. Solution of Equations in Euclidean and Banach Spaces, Third Edition of Solution of Equations and Systems of Equations

H. M. EDWARDS. Riemann's Zeta Function

SAMUEL EILENBERG. Automata, Languages, and Machines: Volume A. *In preparation:* Volume B

MORRIS HIRSCH AND STEPHEN SMALE. Differential Equations, Dynamical Systems, and Linear Algebra

WILHELM MAGNUS. Noneuclidean Tesselations and Their Groups

FRANÇOIS TREVES. Basic Linear Partial Differential Equations

WILLIAM M. BOOTHBY. An Introduction to Differentiable Manifolds and Riemannian Geometry

BRAYTON GRAY. Homotopy Theory: An Introduction to Algebraic Topology

ROBERT A. ADAMS. Sobolev Spaces

JOHN J. BENEDETTO. Spectral Synthesis

D. V. WIDDER. The Heat Equation

IRVING EZRA SEGAL. Mathematical Cosmology and Extragalactic Astronomy

J. DIEUDONNÉ. Treatise on Analysis: Volume II, enlarged and corrected printing; Volume IV

WERNER GREUB, STEPHEN HALPERIN, AND RAY VANSTONE. Connections, Curvature, and Cohomology: Volume III, Cohomology of Principal Bundles and Homogeneous Spaces

In preparation

I. MARTIN ISAACS. Character Theory of Finite Groups

K. D. STROYAN AND W. A. J. LUXEMBURG. Introduction to the Theory of Infinitesimals

D 9
E 0
F 1
G 2
H 3
I 4
J 5